SOCIAL WORK IN MENTAL HEALTH AND SUBSTANCE ABUSE

SOCIAL WORK IN MENTAL HEALTH AND SUBSTANCE ABUSE

Sharon Duca Palmer, CSW, LMSW

School Social Worker, ACLD Kramer Learning Center, Bay Shore, New York; Certified Field Instructor, Adelphi University School of Social Work, Garden City, New York, U.S.A.

Apple Academic Press

Social Work in Mental Health and Substance Abuse

First Published in the Canada, 2011
Apple Academic Press Inc.
3333 Mistwell Crescent
Oakville, ON L6L 0A2
Tel. : (888) 241-2035
Fax: (866) 222-9549
E-mail: info@appleacademicpress.com
www.appleacademicpress.com

The full-color tables, figures, diagrams, and images in this book may be viewed at www.appleacademicpress.com

First issued in paperback 2021

ISBN 13: 978-1-77463-252-9 (pbk)
ISBN 13: 978-1-926692-88-3 (hbk)

Sharon Duca Palmer, CSW, LMSW

Cover Design: Psqua

Library and Archives Canada Cataloguing in Publication Data
CIP Data on file with the Library and Archives Canada

CONTENTS

INTRODUCTION

Social work is a difficult field to operationally define, as it is practiced differently in many settings. It is a very diverse occupation and one that can be practiced in settings such as hospitals, clinics, welfare agencies, schools, and private practices.

The main goal of all social work practice is to assist the client to function at the best of their ability and assess what their needs are. Social workers help clients with problem-solving strategies, such as defining personal goals, focusing on what is necessary to make changes, and helping them through the process.

Social work is a demanding field and is often emotional draining. Many social workers have large caseloads, limited resources for their clients, and often work for relatively low salaries. But the personal rewards can be very satisfying.

The social work profession is committed to promoting social and economic policy though helping to improve people's lives. Research is conducted to improve social services, community development, program evaluation, and public administration. The importance of research in these areas is to examine variables that can be addressed in order to resolve issues. Research can lead to what is called "best practice". By utilizing "best practice", a social worker is engaging clients based on research that is intended to increase successful outcomes.

Social work is one of the most diverse careers available. Most social workers are employed by health care facilities and government agencies. These facilities can include hospitals, mental health clinics, nursing homes, rehabilitation centers, schools, child welfare agencies, and private practice.

Social work's interface with mental health promotion and the treatment of mental illness dates to the earliest roots of our profession. While many social workers provide mental health services in private practice settings, the majority of services are offered in community-based agencies, both public and private, and in hospitals and prisons. Social workers are the largest provider of mental health services, providing more services than all other mental health care providers combined. These workers also often provide services to those who are struggling with substance abuse.

Twenty-first century health issues are complex and multidimensional, requiring innovative responses across professions at all levels of society. Public health social workers work to promote health in hospitals, schools, government agencies and local community-based settings, making connections between prevention and intervention from the individual to the whole population.

In an ideal world, every family would be stable and supportive. Every child would be happy at home and at school. Every elderly person would have a carefree retirement. Yet in reality, many children and families face daunting challenges. For example, single parents struggle to raise kids while working. Teens may become parents before they are ready. Child social workers help kids get back on track so they can lead healthy, happy lives.

Rapid aging populations are expected worldwide. With the rapid growth of this population, social work education and training specializing in older adults and practitioners interested in working with older adults are increasingly in demand. Geriatric social workers typically provide counseling, direct services, care coordination, community planning, and advocacy in an array of organizations including in homes, neighborhoods, hospitals, senior congregate living and nursing facilities. They work with older people, their families and communities, as well as with aging-related policy, and aging research

In whatever subcategory they work, social workers help provide support services to individuals and communities by assessing their needs in order to improve the quality of life and overall well-being. This can lead to positive changes in people's environments, dignity, and self-worth. It can also lead to changes in social policy for those who are vulnerable and oppressed. Social workers change entire communities for the better.

There have been many changes emerging in the social work profession. The uses of the Internet and online counseling have been major trends. Some people are more likely to seek assistance and information first through the use of the Internet. There has also been a strong move for collaborating between professions when providing services in order to offer clients more options for success. Keeping up to date with best practice research, licensing requirements, continuing education, and professional ethics make this an exciting and challenging time to be a social worker!

— Sharon Duca Palmer, CSW, LMSW

Performance Measurement for Co-Occurring Mental Health and Substance Use Disorders

David J. Dausey, Harold A. Pincus and James M. Herrell

ABSTRACT

Background

Co-occurring mental health and substance use disorders (COD) are the norm rather than the exception. It is therefore critical that performance measures are developed to assess the quality of care for individuals with COD irrespective of whether they seek care in mental health systems or substance abuse systems or both.

Methods

We convened an expert panel and asked them to rate a series of structure, process, and outcomes measures for COD using a structured evaluation tool with domains for importance, usefulness, validity, and practicality.

Results

We chose twelve measures that demonstrated promise for future pilot testing and refinement. The criteria that we applied to select these measures included: balance across structure, process, and outcome measures, quantitative ratings from the panelists, narrative comments from the panelists, and evidence the measure had been tested in a similar form elsewhere.

Conclusion

To be successful performance measures need to be developed in such a way that they align with needs of administrators and providers. Policymakers need to work with all stakeholders to establish a concrete agenda for developing, piloting and implementing performance measures that include COD. Future research could begin to consider strategies that increase our ability to use administrative coding in mental health and substance use disorder systems to efficiently capture quality relevant clinical data.

Background

There has been a dramatic growth in recent years in the development of performance measures to monitor and evaluate the quality of medical care [1]. Examples include performance measures developed for the Institute of Medicine's (IOMs) national report card [2], the National Committee for Quality Assurance's (NCQA) Healthcare Effectiveness Data and Information Set (HEDIS) [3], the National Quality Forum's (NQF) "Standardizing Ambulatory Care Performance Measures" project [4], and the American Medical Association's outpatient chronic care clinical performance measures [5].

These measures have been used for everything from internal auditing and quality improvement efforts to benchmarking performance against national averages. Most recently, performance measures are being used to financially reward performance with pay-for-performance initiatives and to distinguish between high and low performing health care systems [6-8]. The substance abuse and mental health fields have long used performance measures such as length of stay, readmission rates, and abstinence during drug treatment. Despite this, the growth of performance measure development in these fields has not kept pace with some other sectors of medicine [9,10]. As with other fields, this interest has been motivated by calls to increase transparency and accountability and thereby improve the overall quality of services delivered [11-13].

The limited numbers of performance measures available tend to focus on either mental health disorders (MHD) [14,15] or substance use disorders (SUD)

[16,17], despite the fact that these disorders have a high probability of co-occurrence. This reflects the segmentation of mental health and substance abuse services into distinct clinical and organizational "silos" [18].

The idea of developing performance measures for co-occurring conditions that cut across these silos is still in its infancy [19]. A notable effort by researchers to develop such measures is the Substance Abuse and Mental Health Services Administration's (SAMHSA) National Outcome Measures (NOMs) project [20]. NOMs comprise ten outcome domains conceptualized to apply to a variety of populations, including persons with COD. Operationalzing NOMs into concrete, measurable performance indicators, however, is still necessary.

SAMSHA contracted with the RAND Corporation in 2006-2007 to begin to conceptualize and develop performance measures for COD. This pilot effort was focused on COD performance measures specifically for SUD settings. However, we discuss how these measures can be easily adapted to be used in MHD settings. The project was designed to rely on the advice of an expert panel to aid in the assessment of a set of candidate COD performance measures. Our goal was to develop a small number of measures that could serve as examples of the types of measures for COD that could be pilot tested and refined for future use. Here we outline the process we used to develop these measures and describe the final set of measures that were developed.

Methods

We convened an expert panel that included a range of stakeholders. Our goal was to include the diverse perspectives of different groups involved in performance measurement in substance abuse and mental health. To this end, we first identified three broad perspectives to include on the panel: (1) state and local perspectives (e.g., individuals with leadership roles related to performance measurement at state and local substance abuse and mental health agencies and national substance abuse and mental health professional associations); (2) health plan perspectives (e.g., individuals who have a role in performance measurement at behavioral health plans); and (3) academic perspectives (e.g., experts from leading academic and research centers who have published on this topic).

We identified a convenience sample of more than thirty experts with backgrounds in at least one of these perspectives and asked them to participate in our panel. Of those contacted, 18 agreed to participate. There was an even distribution of panelists from these perspectives. Based on their career experience, some panelists were able to consider the measures from multiple perspectives. Panelists were asked to respond to a series of quarterly questionnaires over the course of one year. The first two questionnaires asked participants to evaluate the general framework we developed. The remaining questionnaires asked participants to evaluate a three sets of performance measures (structure, process, and outcome) using a systematic data collection form described below. A majority of panelists responded to each questionnaire.

At the start of the project, we asked the panelists to evaluate the basic framework we established to conceptualize, develop, and evaluate our performance measures. The primary conceptual backdrop for the development of the measures was Donabedian's classic triad of structures, processes, and outcomes [21]. We proposed using this framework to categorize our measures.

On the advice of the panelists, we modified this approach by creating subcategories for each part of the triad: (1) structure (e.g., clinician characteristics, clinical information systems, service linkages and financial); (2) process (e.g., detection/identification, assessment, treatment, and service integration/coordination of care) and (3) outcomes (e.g., reduced morbidity, crime and criminal justice, stability in housing, social connectedness, and perception of care).

The subcategories for the structure and process measures were developed from the integration of information derived from the literature, suggestions from the panelists, and suggestions we received during a presentation on this topic at a professional meeting [22]. To ensure consistency with ongoing SAMHSA work in this area, the subcategories for the outcomes measures were derived from the NOMs project.

In developing the performance measures we considered existing work in this area including: (1) performance indicators being developed for the Performance Partnership Grants-PPG [23]; (2) local evaluations of individual Co-Occurring State Incentive Grant (COSIG) projects [24]; (3) quality indicators for substance abuse services developed by the Washington Circle (WC) Group [25]; (4) agency for Healthcare Research and Quality (AHRQ) quality measure clearinghouse [26]; (5) mental health quality indicators developed for the Healthcare Effectiveness Data and Information Set (HEDIS) [3]; and (6) quality indicators identified and/or developed by the Center for Quality Assessment and Improvement in Mental Health (CQAIMH) [27].

Our goal was to develop measures that built on existing work. We started with the assumption that existing performance measures in the substance abuse and mental health fields could be modified to apply to COD. Despite this, we did not want to be constrained to what has already been done. Therefore in addition to adapting existing performance measures to COD, we also developed novel measures that we considered to be of potentially high value.

The candidate measures are meant to serve as exemplars of the types of measures that can be developed. Therefore, some candidate measures specify a particular setting (e.g., inpatient) when they could potentially apply to more than one setting depending on the circumstance. In addition, despite the fact that the measures developed were designed to apply to substance abuse settings, they could easily be adapted to apply to mental health settings.

We asked the panelists to provide feedback on the framework we established for evaluating the measures themselves. We adapted the initial performance measure evaluation tool from one developed by Hermann and colleagues which was provided to the first author electronically (see [27] for more information on Hermann's work). The tool was a simple rating form divided along 6 domains ranging from clinical importance to strategic importance. All of the ratings consisted of 5 point Likert scales. The descriptions of the domains paralleled NCQAs list of desirable attributes for HEDIS measures [3].

We adapted this tool by limiting it to five domains and adapting the descriptions of these domains to apply to COD. We asked the panelists to evaluate our choice of domains and our descriptions of them, and they provided us with feedback on how to improve them. Below we list the final set of domains that we developed based on the feedback from our panelists:

- Importance - Does the measure represent a significant part of the overall quality picture? Does the measure represent an aspect of COD care that is meaningful (e.g., does what is being measured represent a significant quality deficit)?
- Usefulness - Will the use of the measure be helpful in leading to strategies that will result in improvements in patient care? Is the measure likely to have a significant impact on patient care and have the ability to lead to real change in patient outcomes (e.g., is the deficit potentially correctable by an identifiable group)?
- Validity or Scientific Soundness - Is the measure evidence based, is it sensitive to case mix or can it be adjusted to case mix? Can it be gamed or manipulated? Is the measure something that can quantitatively demonstrate changes over time (e.g., if the process is corrected, would the measure be sensitive to detecting this change)?
- Practicality or Feasibility - How practical is it to collect information for the measure (e.g., will data collection require significant costs or be overly burdensome with existing data systems)?
- Overall - What is your overall impression of this measure? Should this measure be included in the core measure set?

When rating the measures, panelists were given detailed descriptions of each measure and asked to rate the measures on each of the domains listed above using a Likert scale where 1 represented strongly agree, 3 represented uncertain, and 5 represented strongly disagree. Space was also provided for panelists to provide narrative comments on each of the measures or to propose additional measures. We also encouraged the panelists to use the space for narrative comments to consider the most appropriate audience for each of the proposed measures. In our initial questionnaire to the panelists we provided several potential examples including: purchasers (public and private), health Plans, agencies/provider organizations, individual providers, consumers, and policymakers.

We developed a total of 36 performance measures broken into structures, processes, and outcomes with at least one measure for each subcategory identified above. We chose a total of twelve measures (four from each category–structure, process, and outcome) that appeared to have the most promise to pilot test. The criteria that we applied to choose the set of twelve measures included: balance across structure, process, and outcome measures, quantitative ratings from the panelists, narrative comments from the panelists, and evidence the measure had been tested in a similar form elsewhere.

Results

Structure Measures

Structure measures tell us whether or not a health care agency or organization has the capacity for process or outcomes measurement (e.g., infrastructure in place that enables quality measurement to occur) and in the case of COD whether or not any integrated or linked services are offered. To measure the quality of COD care, an agency must first have the capacity to deliver integrated or linked services. Structural measures are designed to enable us to access these capacities. Table 1 presents our final refined set of structure measures.

Table 1: Candidate Structure Measures* This table presents all of the candidate structure measures.

Measure Description	Denominator	Numerator	Data Source
1. Assesses the proportion of SUD providers in a SUD specialty care setting who are trained to provide specified mental health care, and who have a certificate, license or some other documentation to demonstrate proficiency.	Total number of SUD providers in a SUD specialty care setting	Total number of providers in the denominator with a certificate, license or other acceptable documentation to prove their competency to provide specified mental health care	Facility data
2. Assesses the proportion of programs in a defined service area (e.g., county, city or state) that report having integrated services (e.g., SUD and MHD services in the same treatment program) or co-located services (e.g., SUD and MHD services in the same location)	Total number of programs in a defined service area	Total number of programs in the denominator that report having integrated or co-located SUD and MHD services	Program records
3. Assesses the proportion of SUD providers in a defined service area (e.g., county, city or state) reporting the ability to bill for MHD services provided to patients.	Total number of SUD providers in a defined service area	Total number of SUD providers in the denominator that report the ability to bill for MHD services provided to patients	SUD provider survey
4. Assesses the proportion of SUD specialty care settings in a defined service area (e.g., county, city or state) that have formal documented referral policies for MHD services.	Total number of SUD specialty care settings in a defined service area	Total number of SUD specialty care settings in the denominator with formal documented referral policies for MHD services	Facility survey

*Measures 1, 3 and 4 can be modified to be used in mental health settings by exchanging SUD for MHD (vice versa). Measure 2 can be applied to both SUD and MHD settings.

Measure 1 assesses the proportion of SUD providers in a SUD specialty care setting who are trained to provide specified mental health care, and who have a certificate, license or some other documentation to demonstrate proficiency. Data for this measure could be collected with a facility survey. A similar measure could be created for MHD providers by assessing the proportion of MHD providers in MHD specialty care settings who have documented proficiency in SUD care.

Measure 2 assesses the proportion of programs in a defined service area (e.g., county, city or state) that report having integrated services (e.g., SUD and MHD services in the same treatment program) or co-located services (e.g., SUD and MHD services in the same location). Program records used to develop this measure could be augmented by the use of standardized instrument measures of fidelity to the integrated services program—see [28] for an example.

Measure 3 assesses the proportion of SUD providers in a defined service area (e.g., county, city or state) reporting the ability to bill for MHD services provided to patients. A similar measure could also be developed for MHD providers providing SUD services where there are also challenges related to billing for COD services.

Measure 4 assesses the proportion of SUD specialty care settings in a defined service area (e.g., county, city or state) that have formal documented referral policies for MHD services. Data for this measure could be collected by a facility survey that could be auditable by facility records.

Measures 1-4 have the potential to be useful for state and local SUD and MHD administrators evaluating the level of integration or the quality of the linkages between SUD and MHD agencies in their jurisdiction.

Process Measures

After we have determined that an agency is capable of measuring performance and that it has some infrastructure in place to deliver integrated or linked services for COD, a logical next step is to consider process measures that examine how care is being delivered. Process measures are a critical link in the chain between structure measures and outcome measures because they enable us to determine whether or not care is being delivered using evidence based standards and guidelines. Without knowing this, it isn't possible to know whether poor outcomes are the result of inadequate program implementation, poor program fidelity, or other factors. Table 2 presents our final refined set of process measures.

Table 2. Candidate Process Measures* This table presents all of the candidate process measures.

Measure Description	Denominator	Numerator	Data Source
5. Assesses the proportion of individuals formally screened for a MHD upon admission to a SUD specialty care setting	Total number of individuals admitted to a SUD specialty care setting	Total number of individuals in the denominator screened for a MHD upon admission	Administrative/claims datasets and medical records
6. Assesses the proportion of individuals that screened positive for COD in a SUD specialty care setting that received a MHD service (or at least one integrated service) within 30 days of screening	Total number of individuals served by a substance abuse agency that screened positive for COD	Total number of individuals in the denominator that received a MHD service (or at least one integrated service) within 30 days of the screening	Administrative/claims datasets or chart review
7. Assesses the proportion of COD with an inpatient or day/night episode (SUD or MHD related) visit that have at least one SUD and one MHD outpatient clinic visit (or one integrated treatment visit) within 30 days of discharge	Total number of COD patients with an inpatient or day/night episode (SUD or MHD related)	Total number of COD patients in the denominator whose medical records indicate that at least one SUD and one MHD outpatient clinic visit (or one integrated treatment visit) within 30 days of discharge	Administrative/claims datasets
8. Assesses the proportion of individuals with COD that were assessed for housing stability	Total number of individuals identified as having a COD	Total number of individuals in the denominator that were assessed for housing stability	Chart review

*Measures 5 and 6 can be modified to be used in mental health settings by exchanging SUD for MHD (vice versa). Measures 7 and 8 can be applied to both SUD and MHD settings.

Measure 5 assesses the proportion of individuals formally screened for a MHD upon admission to a SUD specialty care setting. This measure lends itself to be initially piloted in an inpatient setting where standardized charts are more likely. Collecting data for the measure may require the addition of a field to initial patient intake forms. Importantly, the data collected for this measure could be used for denominator data for other measures.

Measure 6 assesses the proportion of individuals that screened positive for COD in a SUD specialty care setting that received a MHD service (or at least one integrated service) within 30 days of screening. This measure is adapted from measures developed by the Washington Circle Group for substance abuse settings [16] which were pilot tested in six states by state and local substance abuse and/or mental health agencies [17]. The pilot test revealed that state agencies could calculate these types of measures with routinely collected data. A measure such as this might be useful for state-level SUD agencies and might encourage them to develop the structural capacity to measure COD screening.

Measure 7 assesses the proportion of COD with an inpatient or day/night episode (SUD or MHD related) visit that have at least one SUD and one MHD outpatient clinic visit (or one integrated treatment visit) within thirty days of discharge. This measure was adapted from a HEDIS measure for mental illness [29]. The original HEDIS measure was found to be moderately correlated with some but not all similar measures of outpatient performance [30]. Data collection for this measure may require chart reviews in some systems. The denominator data could come from the total number of positive screens found with Measure 5 above.

Measure 8 assesses the proportion of individuals with COD that were assessed for housing stability. A measure such as this could be transitioned into an administrative dataset. The method of identification, however, needs to specify whether it is based on positive screens or based on receipt of both types of services or integrated services. Ultimately it may be possible to develop specific administrative codes to integrate this type of information into administrative datasets.

Measures 5-8 have the potential to be useful to specialty plan administrators to assess the level of integrated or linked services being offered in their plan.

Outcome Measures

Outcomes are the final piece of the performance measurement strategy. Determining what outcomes to measure, however, is not always easy. Different outcomes are important to different stakeholders. For example, patients may be most interested in outcomes that deal with improvements to their quality of life or that improve their functionality while policymakers may be interested in outcomes that reduce crime or inpatient costs. Table 3 presents our final refined set of outcomes measures. All of these measures were based on generic outcomes measures included in NOMs.

Table 3. Candidate Outcomes Measures* This table presents all of the candidate outcome measures.

Measure Description	Denominator	Numerator	Data Source
9. Assesses the proportion of individuals with any MHD discharged from an inpatient or residential SUD specialty care setting with abstinence from drugs and/or alcohol one year after discharge	Total number of individuals discharged from an inpatient or residential SUD specialty care setting with any MHD diagnosis	Total number of individuals in the denominator that report abstinence from drugs and/or alcohol one year after discharge	Patient report and/or laboratory test
10. Assesses the proportion of individuals with any MHD diagnosis discharged from an inpatient or residential SUD specialty setting that move from being unemployed to being employed either part-time or full-time one year after discharge	Total number of individuals discharged from an inpatient or residential SUD specialty care setting with any MHD diagnosis	Total number of individuals in the denominator that move from being unemployed to being employed either part-time or full-time one year after discharge	Patient survey and/or employment records
11. Assesses the proportion of individuals with any MHD diagnosis discharged from an inpatient or residential SUD specialty care setting who report having an episode of incarceration within 6 months of discharge	Total number of individuals discharged from an inpatient or residential SUD specialty care setting with any MHD diagnosis	Total number of individuals in the denominator reporting an episode of incarceration within 6 months of discharge	Patient survey and/or criminal justice system data
12. Assesses the proportion of individuals receiving care in a SUD specialty care setting with any MHD diagnosis who report improved satisfaction with their care as measured by a standardized instrument after 6 months of treatment	Total number of individuals receiving care in a SUD specialty care setting with any MHD diagnosis	Total number of individuals in the denominator who report improved satisfaction with their care after 6 months of treatment	Patient survey

*All of these measures can be modified to be used in mental health settings by exchanging SUD for MHD (vice versa).

Measure 9 assesses the proportion of individuals with any MHD discharged from an inpatient or residential SUD specialty care setting with abstinence from drugs and/or alcohol one year after discharge. The data for this measure could come from patient report and/or laboratory tests. This measure focuses on inpatient care; however, it is possible for a similar measure to be developed for outpatient care.

Measure 10 assesses the proportion of individuals with any MHD diagnosis discharged from an inpatient or residential SUD specialty setting that move from being unemployed to being employed either part-time or full-time one year after discharge. The data from this measure could come from patient report and/or employment records. As with Measure 9, a similar measure could be developed for outpatient settings.

Measure 11 assesses the proportion of individuals with any MHD diagnosis discharged from an inpatient or residential SUD specialty care setting who report having an episode of incarceration within 6 months of discharge. Data for this measure could come from patient report and/or criminal justice system data. As with Measures 9-10, a similar measure for outpatient settings could also be developed.

Measure 12 assesses the proportion of individuals receiving care in a SUD specialty care setting with any MHD diagnosis who report improved satisfaction with their care as measured by a standardized instrument after 6 months of treatment. Data for this measure could come from satisfaction measures included on patient surveys. The respondent pool for these surveys could be determined by claims data or a checkbox for self-report of a MHD.

Measures 9-12 have the potential to be useful to SUD and MHD providers to assess how well they are providing integrated or linked SUD and MHD services.

Discussion

Developing reliable and valid performance measures for COD that are of high value while remaining feasible to implement is a significant challenge. This project sought to take a step forward in the development of such measures by developing a framework to conceptualize these measures and by considering a small set of measures for substance abuse settings evaluated through expert consensus. The model introduced by this project for initially developing performance measures for COD is a logical first step and could be expanded and replicated elsewhere.

Because of the qualitative nature of our analysis, we do not present the raw data from our analysis of the expert panelist feedback. This feedback was provided in the form of ratings using Likert scales in combination with written qualitative feedback for each measure. It is important to note, however, that overall our panelists rated outcomes measures as more important than structure and process measures. This likely reflects the general belief that outcomes are the ultimate aspect of quality that we are trying to assess in any performance measurement system. Despite this, in choosing our final set of measures, we purposely included an equal number of structure, process, and outcomes. We agree with the importance of outcomes measures; however, we argue that these measures need to be part of an integrated package of measures with the intent for full implementation over time. To interpret outcomes and to use outcomes to improve service, measures of structure and process are essential.

A tension in performance measurement is the extent to which measures can be developed using existing administrative data systems versus those that require additional data collection. Using administrative data is less costly and involves less effort; however, these measures tend to be "black box" measures that don't really allow us to directly assess what we want to measure. Developing new data systems or additional data collection applets in old data systems can be costly and if done without proper foresight can yield limited benefits. It can be argued, however, that without considering new or modified data systems that we will never be able to measure quality adequately.

For many psychosocial problems (e.g., homelessness, unemployment, family interventions, legal problems), without having a simple and efficient way of identifying them (i.e., identifying who needs help) it is difficult to develop performance measures that do not require extensive chart reviews. This is less of a problem with developing performance measures so much as it is a problem with making the data collection for these measures feasible. A coding system to capture this information that could be integrated into the clinical record is one way to solve this problem [31].

This project has several limitations. In choosing the measures, we tried to achieve a balance between attributes with a particular emphasis on practicality. There are many other measures that could be developed. The measures presented here are intended to serve as exemplars of the types of measures that could be developed. They also serve as measures that are potentially promising to pilot test.

The setting in which each measure is most applicable varies (e.g., health plans, SUD organizations, providers). Our emphasis was to focus on settings that were most likely to have data. That data for some of these measures could become unstable for smaller organizations. Nonetheless the measure sets a clear goal and can aid even these small organizations in moving towards that goal.

Several of the measures that we developed focused on inpatient care settings. This emphasis is not meant to suggest that outpatient care is not important. More outpatient measures will need to be developed over time. Inpatient measures; however, provide us with an environment where we have the most control and provide us with a date to start counting. In addition, appropriate follow-up after inpatient care is a high priority.

These measures and others that may be developed in the future will need to be continually refined as we gain an understanding of their psychometric and statistical properties. Refinements will need to include weighting procedures and other statistical adjustments. For example, Measure 3 assesses the proportion of SUD providers within a geographical area that are able to bill for MHD services. Compared to community based SUD providers, hospital based SUD providers may be more likely to be able to bill for MHD services while at the same less likely to see as many clients. Thus areas with higher proportions of hospital based providers compared to community based providers could score higher on this measure while treating fewer COD clients than areas with the reverse. Strategies to appropriately adjust for caseload will be needed to solve these types of problems.

This project raises several avenues for future research in this area. A logical next step is to pilot test these measures (or other measures that are developed) in real settings. These pilot tests could take on one of two forms: pilot tests of measures that lend themselves to existing administrative datasets versus pilot tests of measures that require additional data collection. Existing administrative datasets of one or more health care organizations can used to test the feasibility of the current specifications of the measures. Modifications can be made to these specifications based on what is learned. It may turn out that some of the measures that initially seemed feasible with administrative data are less amenable to this format than originally thought. Measures that involve data not contained in administrative or claims datasets will require additional data collection. These measures can be pilot tested in practice care settings through surveys of patients, providers, and family members.

Conclusion

Having clear processes for developing, vetting and piloting performance measures affords us an opportunity to think about what we want to measure and contrast that with what we are able to measure with existing data systems and infrastructure. Policymakers have a vast array of considerations to take into account when choosing performance measures. They must consider the implementation of new performance measures against the backdrop of scarce resources and competing priorities. New performance measures are critical but may not be possible without additional resources. It is important that as this process moves forward that all stakeholders in the field have an opportunity to review and comment on new performance measures being considered. To be successful performance measures need to be developed in such a way that they align with needs of administrators and providers. Policymakers need to work with all stakeholders to establish a concrete agenda for developing, piloting and implementing performance measures that include COD.

Competing Interests

The authors declare that they have no competing interests.

Authors' Contributions

DJD and HAP developed the measures and conducted the background research. DJD took primary responsibility for writing all sections of the paper. JMH supervised the work and helped to write the background and conclusion sections of the paper. All authors read and approved the final manuscript.

Acknowledgements

This work was jointly supported by a contract from the Substance Abuse and Mental Health Services Administration (SAMHSA)/Center for Substance Abuse Treatment (CSAT) and a grant from the Robert Wood Johnson Foundation Substance Abuse Policy Research Program (SAPRP). We would like to thank the participants of our expert panel for their comments that helped to shape this paper.

References

1. Clancy C: The performance of performance measurement. Health Serv Res 2007, 42:1797–801.
2. Hurtado P, Swift EK, Corrigan JM: Envisioning the national health care quality report. Washington, DC: National Academy Press; 2001.
3. National Committee for Quality Assurance [http://web.ncqa.org/tabid/59/Default.aspx].
4. National Quality Forum [http://www.ahrq.gov/qual/aqastart.htm].
5. Keyser DJ, Dembosky JW, Kmetik K, Antman MS, Sirio C, Farley DO: Using health information technology-related performance measures and tools to improve chronic care. Jt Comm J Qual Patient Saf 2009, 35:248–55.

6. Kahn CN, Ault T, Isenstein H, Potetz L, Van Gelder S: Snapshot of hospital quality reporting and pay-for-performance under Medicare. Health Aff 2006, 25:148–62.
7. Lee TH: Pay for performance, version 2.0? N Engl J Med 2007, 357:531–3.
8. Rosenthal MB, Landon BE, Howitt K, Song HR, Epstein AM: Climbing up the pay-for-performance learning curve: where are the early adopters now? Health Aff 2007, 26:1674–82.
9. Ganju V: Mental health quality and accountability: the role of evidence-based practices and performance measurement. Adm Policy Ment Health 2006, 33:659–65.
10. Levy Merrick E, Garnick DW, Horgan CM, Hodgkin D: Quality measurement and accountability for substance abuse and mental health services in managed care organizations. Med Care 2002, 40:1238–48.
11. Hunsicker RJ: Time to be transparent and accountable. Behav Healthc 2007, 27:48.
12. McLellan AT, Chalk M, Bartlett J: Outcomes, performance, and quality: what's the difference? J Substan Abuse Treat 2007, 32:331–40.
13. Pincus HA, Page AE, Druss B, Appelbaum PS, Gottlieb G, England MJ: Can psychiatry cross the quality chasm? Improving the quality of health care for mental and substance use conditions. Am J Psychiatry 2007, 164:712–9.
14. Bramesfeld A, Klippel U, Seidel G, Schwartz FW, Dierks ML: How do patients expect the mental health service system to act? Testing the WHO responsiveness concept for its appropriateness in mental health care. Soc Sci Med 2007, 65:880–9.
15. Dausey DJ, Rosenheck RA, Lehman AF: Preadmission Care as a New Mental Health Performance Indicator. Psychiatr Serv 2002, 53:1451–5.
16. McCorry F, Garnick DW, Bartlett J, Cotter F, Chalk M: Developing performance measures for alcohol and other drug services in managed care plans. Washington Circle Group. Jt Comm J Qual Improv 2000, 26:633–43.
17. Garnick DW, Lee MT, Horgan CM, Acevedo A, Washington Circle Public Sector Workgroup: Adapting Washington Circle performance measures for public sector substance abuse treatment systems. J Substan Abuse Treat 2009, 36:265–77.
18. Horvitz-Lennon M, Kilbourne AM, Pincus HA: From Silos to Bridges: Meeting the General Health Care Needs of Adults with Severe Mental Illnesses. Health Aff 2006, 25:659–69.

19. Kilbourne AM, Fullerton CA, Dausey DJ, Pincus HA, Hermann RC: A Framework for Measuring Quality across Silos: The Case of Mental Disorders and Co-occurring Conditions. Qual Saf Health Care 2009, in press.
20. Substance Abuse and Mental Health Services Administration [http://www.oas.samhsa.gov/NOMsCoOccur2k6.pdf].
21. Donabedian A: Explorations in Quality Assessment and Monitoring: The Definition of Quality and Approaches to its Assessment. Ann Arbor: Health Administration Press; 1980.
22. Annual Meeting of the Depression and Primary Care Program [http://www.cqaimh.org/DPC%20Presentation.ppt].
23. National Association of State Alcohol/Drug Abuse Directors [http://www.nasadad.org/index.php?doc_id=161].
24. Dausey DJ, Pincus HA, Herrell JM, Rickards L: States' early experience in improving systems-level care for persons with co-occurring disorders. Psychiatr Serv 2007, 58:903–5.
25. Washington Circle Group [http://www.washingtoncircle.org/].
26. Agency for Healthcare Research and Quality [http://www.qualitymeasures.ahrq.gov/].
27. Center for Quality Assessment and Improvement in Mental Health [http://www.cqaimh.org/index.html].
28. McHugo GJ, Drake RE, Whitley R, Bond GR, Campbell K, Rapp CA, Goldman HH, Lutz WJ, Finnerty MT: Fidelity outcomes in the National Implementing Evidence-Based Practices Project. Psychiatr Serv 2007, 58:1279–84.
29. National Quality Measures Clearinghouse [http://www.qualitymeasures.ahrq.gov/].
30. Druss B, Rosenheck R: Evaluation of the HEDIS measure of behavioral health care quality. Health Plan Employer Data and Information Set. Psychiatr Serv 1997, 48:71–75.
31. First MB, Pincus HA, Schoenbaum M: Issues for DSM-V: Adding Problem Codes to Facilitate Assessment of Quality of Care. Am J Psychiatry 2009, 166:11–3.

Methamphetamine Use and Rates of Incarceration Among Street-Involved Youth in a Canadian Setting: A Cross-Sectional Analysis

M-J. Milloy, Thomas Kerr, Jane Buxton, Julio Montaner and Evan Wood

ABSTRACT

Background

Given concerns over rising use of methamphetamine, especially among street-involved youth, and the links between exposure to the correctional system and the production of drug-related harm, we sought to assess the relationship between ever using methamphetamine and reporting ever being incarcerated in the At-Risk Youth Survey (ARYS) in Vancouver, Canada.

Methods

The relationship between ever being imprisoned and ever using methamphetamine was estimated using a multivariate logistic regression analysis while also considering potentially confounding secondary demographic, social and behavioural variables.

Results

Of the 478 youth recruited into ARYS between September 2005 and October 2006, 385 (80.5%) reported ever being incarcerated overnight or longer. In the multivariate model, methamphetamine use was independently associated with ever being incarcerated (Adjusted Odds Ratio: 1.79, 95% Confidence Interval [CI]: 1.03–3.13).

Conclusion

Incarceration was very common in this cohort and strongly linked with ever using methamphetamine. This finding is of concern and, along with the previously identified risks of drug-related harm associated with incarceration, supports the development of novel public policy, such as community-based drug treatment, to address the use of methamphetamine among street youth.

Background

The use of methamphetamine in Western settings is of increasing concern [1,2], especially among street-involved youth [3,4], a vulnerable population already burdened by high levels of morbidity and mortality [5,6]. According to the United Nations Office on Drugs and Crime, methamphetamine now constitutes the second most commonly used illicit drug internationally, second only to marijuana [7].

For older drug users, especially those who use injection drugs (IDU), the dynamics linking drug use, marginalisation and imprisonment are well described [8-10]. Arrest and imprisonment is a common experience, with a history of incarceration reported by at least 75% of participants in community-recruited samples of IDU in Europe [11], Thailand [12] and the United States [13]. Incarceration may be a risk factor for drug related harm among IDU, since exposure to correctional environments has consistently been associated with an increased likelihood of HIV risk behavior and HIV infection [14,15] as well as increased risk of fatal overdose upon release [16].

Sparked by the growing use of methamphetamine and concerns over links to initiation of injection drug use [17], we have previously reported that over 75% of participants in a local cohort of street-involved youth said they had previously used methamphetamine [4]; 25% of all injection initiation experiences involved methamphetamine [4]; and 13% of local overdose events among homeless youth involve the use of methamphetamine [18]. Vancouver is the site of an explosive outbreak of HIV among IDU with current prevalence estimated at 20% [19]; approximately 3% of local street youth are estimated to be HIV-seropositive [20]. Since exposure to the criminal justice system through arrest and incarceration may actually increase drug-related harms [15], we conducted the present study to determine the prevalence of incarceration in a cohort of community-recruited street youth and investigate its relationship with the use of methamphetamine.

Methods

The At-Risk Youth Study (ARYS) is a prospective cohort of street-involved youth in Vancouver, Canada, that has been described in detail previously [17]. In brief, snowball sampling and street-based outreach were used in an effort to derive a representative sample of street-involved drug using youth. Individuals were eligible for inclusion if they were aged 14 to 26 years old at the baseline interview and had used illegal drugs other than cannabinoids in the previous 30 days. At baseline and every six-month follow-up, participants answer an interviewer-administered questionnaire, are examined by a nurse and provide blood samples for serologic testing. The ARYS study has been reviewed and approved by the University of British Columbia/Providence Research Ethics Board.

For the present analyses, the outcome of interest was reporting ever being incarcerated, or answering "yes" to the question: "Have you ever been in detention, prison, the drunk tank or jail overnight or longer?" The primary explanatory variable of interest was reporting ever using methamphetamine. First, we compared individuals reporting incarceration with those reporting never incarcerated using individual-, social- and structural-level factors we hypothesised could be associated with both the outcome of interest and primary explanatory variable. These secondary explanatory variables included: gender; age; ethnicity (Non-aboriginal vs. aboriginal); education level (< high school vs. ≥ high school); history of foster care (yes vs. no); history of ER use (yes vs. no); hepatitis C virus (HCV) seropositivity (yes vs. no); ever involved in the sex-trade (yes vs. no); ever diagnosed with a mental illness (yes vs. no); ever dealing drugs (yes vs. no); ever being sexually or physically abused (yes vs. no); ever using crack cocaine (yes vs. no); ever using powder cocaine (yes vs. no); ever injecting heroin (yes vs. no); ever using cannibinoids (i.e., marijuana, hashish) (yes vs. no). All drug use variables referred to any prior use.

For univariate analyses, we used Pearson's $\chi 2$ test (dichotomous variables) and the Mann-Whitney test (continuous variables) to compare individuals reporting the outcome versus others by the primary and secondary explanatory variables. To fit the multivariate model, we employed a backwards selection procedure we have used previously [21,22]. After beginning with a full model with all covariates included, we fit reduced models, each with one unique secondary explanatory variable removed, and observed in each model the relative change in the coefficient for the term for methamphetamine in the regression equation. We identified the reduced model with the smallest absolute relative change in the methamphetamine coefficient and removed its missing secondary variable from further consideration. The objective of this step is to remove variables with relatively less effect on the value of the coefficient for methamphetamine and, with each step, to preserve variables in the analysis with greater infuence on the value of the methamphetamine coefficient in multivariate analysis. We continued this iterative process until the smallest relative change in the methamphetamine coefficient exceeded 5% of the value of the coefficient. We then fit a final model including methamphetamine use and all remaining secondary explanatory variables as terms in the regression equation.

All statistical analyses were performed in R version 2.6.1 (R Foundation for Statistical Computing, Vienna, Austria). All p-values are two-sided.

Results

Between September 2005 and October 2006, 478 individuals were recruited into the ARYS cohort, of whom 132 (27.6%) were female, 120 (25.1%) reported Aboriginal ancestry and 329 (68.8%) were Caucasian. At the baseline interview, the median age was 22.0 (Interquartile Range [IQR]: 20.0–23.9).

Of the 478 participants, 385 (80.5%) reported ever being incarcerated. As shown in Table 1, social and demographic characteristics associated with incarceration in univariate analyses were: older age (Odds Ratio [OR]: 1.23, 95% Confidence Interval [95% CI]: 1.17–1.28, $p < 0.001$); having less than a high school education (OR: 1.66, 95% CI: 1.04–2.66, $\chi^2 = 4.07$ [df = 1], $p = 0.032$); and ever being a victim of abuse (OR: 2.10, 95% CI: 1.32–3.34, $\chi^2 = 9.24$ [df = 1], $p = 0.002$). Female gender was inversely associated with having a history of incarceration (OR: 0.19, 95% CI: 0.12–0.31, $\chi^2 = 48.03$ [df = 1], $p < 0.001$). Behavioural and drug-using variables associated with a history of incarceration are shown in Table 2 and included: methamphetamine use (OR: 2.45, 95% CI: 1.53–3.90, $\chi^2 = 13.53$ [df = 1], $p < 0.001$); crack use (OR: 3.08, 95% CI: 1.89–5.03, $\chi^2 = 20.12$ [df = 1], $p < 0.001$); cocaine use (OR: 2.49, 95% CI: 1.33–4.66, $\chi^2 = 8.02$ [df = 1], $p = 0.003$); and drug dealing (OR: 3.19, 95% CI: 1.97–5.19, $\chi^2 = 22.03$ [df = 1], $p < 0.001$).

Table 1. Univariate analyses of social and demographic characteristics associated with reporting ever being incarcerated in ARYS (n = 478)

Characteristic	Ever incarcerated No (%)	Ever incarcerated Yes (%)	Odds Ratio	95% Confidence Interval	*p*-value[1]
Age (df = 476)					
Median (IQR)	20.8 (17.7 – 23.4)	22.4 (18.4 – 26.3)	**1.23**	**1.17 – 1.28**	**< 0.001**
Gender (df = 1)					
Male	40 (43.0)	306 (79.5)			
Female	53 (57.0)	79 (20.5)	**0.19**	**0.12 – 0.31**	**< 0.001**
Ethnicity (df = 1)					
Non-Aboriginal	76 (81.7)	282 (73.2)			
Aboriginal	17 (18.2)	103 (26.8)	1.63	0.92 – 2.89	0.091
Education (df = 1)					
≥ High school	38 (40.9)	113 (29.4)			
< High school	55 (59.1)	272 (70.6)	**1.66**	**1.04 – 2.66**	**0.032**
Foster care[2] (df = 1)					
No	52 (55.9)	183 (47.5)			
Yes	41 (44.1)	202 (52.5)	1.40	0.89 – 2.21	0.147
HCV status (df = 1)					
Negative	77 (82.8)	339 (88.1)			
Positive	16 (17.2)	46 (11.9)	0.65	0.35 – 1.21	0.176
Mental illness[2] (df = 1)					
No	59 (63.4)	225 (58.4)			
Yes	34 (36.6)	160 (41.6)	1.23	0.77 – 1.97	0.378
Victim of abuse[2] (df = 1)					
No	58 (62.4)	170 (44.2)			
Yes	35 (37.6)	215 (55.8)	**2.10**	**1.32 – 3.34**	**0.002**
ER use[2] (df = 1)					
No	61 (65.6)	225 (58.4)			
Yes	32 (34.4)	160 (41.6)	1.36	0.84 – 2.18	0.239

1. *p*-values based on χ-square tests of difference (for categorical variables) and the Mann-Whitney test (for continuous)
2. Refers to any instance in the past

Table 2. Univariate analyses of behavioural and drug-using characteristics associated with reporting ever being incarcerated in ARYS (n = 478)

Characteristic	Ever incarcerated No (%)	Ever incarcerated Yes (%)	Odds Ratio	95% Confidence Interval	*p*-value[1]
Methamphetamine use[2] (df = 1)					
No	42 (45.2)	97 (25.2)			
Yes	51 (54.8)	288 (74.8)	**2.45**	**1.53 – 3.90**	**< 0.001**
Crack use[2] (df = 1)					
No	37 (39.8)	68 (17.7)			
Yes	56 (60.2)	317 (82.3)	**3.08**	**1.89 – 5.03**	**< 0.001**
Heroin injection[2] (df = 1)					
No	74 (79.7)	269 (69.9)			
Yes	19 (20.4)	116 (30.1)	1.68	0.97 – 2.90	0.062
Cocaine use[2] (df = 1)					
No	80 (86.0)	274 (71.1)			
Yes	13 (14.0)	111 (28.9)	**2.49**	**1.33 – 4.66**	**0.003**
Cannabinoid use[2] (df = 1)					
No	8 (8.6)	19 (4.9)			
Yes	85 (91.4)	366 (93.1)	1.81	0.76 – 4.28	0.169
Drug dealing[2] (df = 1)					
No	39 (41.9)	71 (18.4)			
Yes	54 (58.1)	314 (81.6)	**3.19**	**1.97 – 5.19**	**< 0.001**
Sex trade[2] (df = 1)					
No	77 (82.8)	305 (79.2)			
Yes	16 (17.2)	80 (20.8)	1.26	0.70 – 2.28	0.440

1. *p*-value based on results of χ-square test of difference
2. Refers to any time in the past

Results from the final multivariate logistic regression model are displayed in Table 3. The primary explanatory variable, previous use of methamphetamine, was independently associated with ever being incarcerated in a model which included foster care, female gender, Aboriginal ethnicity and crack use. Correlation between the explanatory variables was moderate, ranging from 0.00 to 0.35.

Table 3. Multivariate logistic regression analysis of primary and secondary factors associated with reporting ever being incarcerated in ARYS (n = 478)

Characteristic	Adjusted Odds Ratio	95% Confidence Interval	p-value[2]
Methamphetamine use[1] (yes vs. no)	**1.79**	**1.03 – 3.13**	**0.041**
Foster care[1] (yes vs. no)	1.58	0.94 – 2.65	0.081
Gender (Female vs. male)	0.17	0.10 – 0.28	< 0.001
Ethnicity (Aboriginal vs. non-Aboriginal)	1.69	0.89 – 3.18	0.107
Crack use[1] (yes vs. no)	2.45	1.38 – 4.32	0.002

1. Refers to any time in the past

Discussion

In this survey of street-involved youth in Vancouver, Canada, we observed a high level of both ever being incarcerated and ever using methamphetamine. The level of incarceration observed in this sample (80.5%) is substantially higher than other estimates in surveys of street-involved youth. In 2004, a multi-site cross-sectional study of 1733 Canadian street youth reported 784 (45.2%) had been in jail [23]. A similar level was reported by 536 homeless youth in Portland, Oregon [24]. In our setting, this level of incarceration is higher (80.5% vs. 59.4%) than that observed in a cohort of adult IDU recruited from a local harm reduction facility [15]. Reasons for this heightened level might include, proximally, the prevalence of high-intensity drug use and involvement in the sex trade; and, ultimately, social and structural factors including a dearth of affordable housing and ordnances targeting homeless individuals [25,26].

Although several street youth surveys include contact with the criminal justice system as an explanatory covariate [23,27,28], we are unaware of any study that identifies the factors associated with incarceration among street-involved youth. In the present study, we found methamphetamine use to be independently associated with ever being incarcerated after adjustment for a number of possible social, demographic and behavioural confounders. Since it is not possible to resolve the temporal relationship between the dependent and primary explanatory variable in a cross-sectional analysis, we hypothesise the association is most likely the result of methamphetamine use, and the means required to support it (e.g., sex trade involvement and other criminal activity) increasing the visibility of street youth to police, elevating the risk of arrest and imprisonment. However, the possibility that methamphetamine use is a sequelae of imprisonment for some individuals cannot be excluded. Numerous studies report a shift to higher-intensity drug use, for example the initiation of drug use by injection, upon incarceration [12,29,30]. Similarly, in a sample of 569 street-involved young men who have sex with men in New York City, contact with the criminal justice system was most often found to precede beginning to use drugs such as heroin, cocaine and speed as well as involvement in the sex trade [30]. In a detailed qualitative analysis, Vancouver street-involved youth described the multiple ways methamphetamine use helped them cope with their social and environmental circumstances, including mediating social contacts, maintaining vigilance over themselves and their possessions, and avoiding the use of psychiatric medications [29].

Regardless of whether methamphetamine use is a predictor or sequelae of incarceration, the strong independent association observed between its use and imprisonment in this analysis is cause for concern. As a result of the persistence of drug use by many prisoners [31] alongside the lack of harm reduction and addiction treatment opportunities within correctional environments [31], exposure to correctional environments has been linked to a higher risk for infection with blood-borne pathogens, including HIV, in this setting [15] as others [32,33]. Thus, the frequent imprisonment of street youth who inject methamphetamine could help sustain viral transmission in this population. Although future work should investigate the relationship between contact with police, courts and jails and intake into alcohol and drug treatment programmes for young drug users, the brief sentences typically served by those designated young offenders suggests little rehabilitative care is available [34]. These factors support the development of novel public policies to address methamphetamine use. We recognise that a substantial segment of policymakers as well as the general public supports punitive sanctions for illicit drug use as a signal of social disapproval as well as a disincentive for current or future use. However, we note that little empiric evidence exists of the effectiveness of this approach on either the individual or population level despite the investment of significant public funds [35]. Thus, our findings add support to calls for new policy approaches to curb illicit drug use among members of the population, for example community diversion or expanded access to drug treatment. Some new programmes to address methamphetamine use, especially in the United States, have been developed, including education and public awareness and precursor regulation [36,37]. These initiatives should be rigourously evaluated before being applied to a vulnerable population.

We also observed a high prevalence of ever using crack cocaine in this cohort, with 78.0% of participants reporting ever using the drug. In the univariate analysis, crack cocaine use was strongly associated with ever being incarcerated ($p < 0.001$). While the effect measures of secondary adjusting variables included in confounding models should be interpreted with caution, it is clear that there is a strong and likely independent effect of crack cocaine use increasing the likelihood of incarceration. The link between high-intensity cocaine use and a greater likelihood of drug-related harms, including incarceration, has been well described in this and other settings. Recently, we reported a high level of crack use in this cohort strongly linked with homelessness [38]. Previous research from Vancouver determined that stimulant use, including cocaine and methamphetamine, helps individuals cope with the immediate rigours of street-involved life, including diminishing feelings of hunger, improving wakefullness and awareness and reducing boredom [29,38].

This analysis has some limitations which should be addressed. As random sampling methods could not be employed due to a lack of voters' lists or other registries, findings from this population of street-involved youth might not be generalisable to the entire local street youth population or other settings. However, it is noteworthy the demographic composition of ARYS is similar to other street-youth samples in Vancouver [3,39]. Second, several measures rely on self-report; thus, social desirability bias might have led to an underestimate of the prevalence of some variables. However, we do not believe any bias was differentially reported by history of incarceration. Finally, we were unable to consider the effect of different durations or locations of incarceration nor did we gather information on the age at first incarceration; also, the cohort contains individuals possibly exposed to either youth detention centres, adult facilities, or both. Future research should consider the effect of these modifiers on drug use patterns and other concerns.

Conclusion

To conclude, this is the first study to describe such high rates of incarceration among street involved youth and to explore risk factors for incarceration among this population. In multivariate regression analysis including several possible confounders, reporting a history of incarceration was strongly associated with ever using methamphetamine. Given the rising prevalence of methamphetamine use in this area as others, and the elevated risk for drug-related harms including HIV infection associated with exposure to correctional environments, these findings support the development of new public policy to support the health of drug-using and street-involved youth, and the exploration of community diversion programs (e.g. addiction treatment) to avoid the high rates of incarceration among this population.

Competing Interests

M-JM, JB, EW and TK declare they have no competing interests. JM has received educational grants from, served as an ad hoc adviser to or spoken at various events sponsored by Abbott Laboratories, Agouron Pharmaceuticals Inc., Boehringer Ingelheim Pharmaceuticals Inc., Borean Pharma AS, Bristol-Myers Squibb, DuPont Pharma, Gilead Sciences, GlaxoSmithKline, Hoffmann-La Roche, Immune Response Corporation, Incyte, Janssen-Ortho Inc., Kucera Pharmaceutical Company, Merck Frosst Laboratories, Pfizer Canada Inc., Sanofi Pasteur, Shire Biochem Inc., Tibotec Pharmaceuticals Ltd. and Trimeris Inc.

Authors' Contributions

M-JM and EW conceived the study. EW, TK and M-JM designed the analysis; M-JM performed the statistical procedures. M-JM wrote the manuscript and incorporated all suggestions. JB provided information and edited a draft of the manuscript. JM contributed to the conception and design of the analysis, interpretation of the data and drafting of the report. All authors approved the version to be published.

Acknowledgements

We would particularly like to thank the ARYS participants for their willingness to be included in the study, as well as current and past ARYS investigators and staff. We would specifically like to thank Deborah Graham, Tricia Collingham, Caitlin Johnston, Steve Kain, and Calvin Lai and Leslie Rae for their research and administrative assistance. The study was supported by the US National Institutes of Health and the Canadian Institutes of Health Research. Thomas Kerr is supported by the Michael Smith Foundation for Health Research and the Canadian Institutes of Health Research.

References

1. Buxton JA, Dove NA: The burden and management of crystal meth use. CMAJ 2008, 178:1537–1539.
2. Degenhardt L, Roxburgh A, Black E, Bruno R, Campbell G, Kinner S, Fetherston J: The epidemiology of methamphetamine use and harm in Australia. Drug Alcohol Rev 2008, 27:243–252.
3. Martin I, Lampinen TM, McGhee D: Methamphetamine use among marginalized youth in British Columbia. Can J Public Health 2006, 97:320–324.
4. Wood E, Stoltz JA, Zhang R, Strathdee SA, Montaner JS, Kerr T: Circumstances of first crystal methamphetamine use and initiation of injection drug use among high-risk youth. Drug Alcohol Rev 2008, 27:270–276.

5. Ensign J, Gittelsohn J: Health and access to care: perspectives of homeless youth in Baltimore City, USA. Soc Sci Med 1998, 47:2087–2099.
6. Roy E, Haley N, Leclerc P, Sochanski B, Boudreau JF, Boivin JF: Mortality in a cohort of street youth in Montreal. Jama 2004, 292:569–574.
7. World Drug Report 2008 Vienna, Austria: United Nations Office of Drugs and Crime; 2008.
8. Burris S, Blankenship KM, Donoghoe M, Sherman S, Vernick JS, Case P, Lazzarini Z, Koester S: Addressing the "risk environment" for injection drug users: the mysterious case of the missing cop. Milbank Q 2004, 82:125–156.
9. Drucker E: Drug prohibition and public health: 25 years of evidence. Public Health Rep 1999, 114:14–29.
10. Kerr T, Small W, Wood E: The public health and social impacts of drug market enforcement: A review of the evidence. International journal of drug policy 2005, 16:210–220.
11. Ball A: Multi-centre study on drug injecting and risk of HIV infection: a report prepared on behalf of the International Collaborative Group for the World Health Organization Programme on Substance Abuse. Geneva, Switzerland: World Health Organization; 1995.
12. Thaisri H, Lerwitworapong J, Vongsheree S, Sawanpanyalert P, Chadbanchachai C, Rojanawiwat A, Kongpromsook W, Paungtubtim W, Sri-ngam P, Jaisue R: HIV infection and risk factors among Bangkok prisoners, Thailand: a prospective cohort study. BMC Infect Dis 2003, 3:25.
13. Clarke JG, Stein MD, Hanna L, Sobota M, Rich J: Active and Former Injection Drug Users Report of HIV Risk Behaviors During Periods of Incarceration. Substance abuse 2001, 22:209–216.
14. Buavirat A, Page-Shafer K, van Griensven GJ, Mandel JS, Evans J, Chuaratanaphong J, Chiamwongpat S, Sacks R, Moss A: Risk of prevalent HIV infection associated with incarceration among injecting drug users in Bangkok, Thailand: case-control study. Bmj 2003, 326:308.
15. Milloy MJ, Wood E, Small W, Tyndall M, Lai C, Montaner J, Kerr T: Incarceration experiences in a cohort of active injection drug users. Drug Alcohol Rev 2008, in press.
16. Binswanger IA, Stern MF, Deyo RA, Heagerty PJ, Cheadle A, Elmore JG, Koepsell TD: Release from prison–a high risk of death for former inmates. N Engl J Med 2007, 356:157–165.
17. Wood E, Stoltz JA, Montaner JS, Kerr T: Evaluating methamphetamine use and risks of injection initiation among street youth: the ARYS study. Harm Reduct J 2006, 3:18.

18. Werb D, Kerr T, Lai C, Montaner J, Wood E: Nonfatal overdose among a cohort of street-involved youth. J Adolesc Health 2008, 42:303–306.

19. Wood E, Kerr T, Marshall BD, Li K, Zhang R, Hogg RS, Harrigan PR, Montaner JS: Longitudinal community plasma HIV-1 RNA concentrations and incidence of HIV-1 among injecting drug users: prospective cohort study. Bmj 2009, 338:b1649.

20. Marshall BD, Kerr T, Livingstone C, Li K, Montaner JS, Wood E: High prevalence of HIV infection among homeless and street-involved Aboriginal youth in a Canadian setting. Harm Reduct J 2008, 5:35.

21. Lima VD, Geller J, Bangsberg DR, Patterson TL, Daniel M, Kerr T, Montaner JS, Hogg RS: The effect of adherence on the association between depressive symptoms and mortality among HIV-infected individuals first initiating HAART. Aids 2007, 21:1175–1183.

22. Wood E, Hogg RS, Lima VD, Kerr T, Yip B, Marshall BD, Montaner JS: Highly active antiretroviral therapy and survival in HIV-infected injection drug users. Jama 2008, 300:550–554.

23. Shields SA, Wong T, Mann J, Jolly AM, Haase D, Mahaffey S, Moses S, Morin M, Patrick DM, Predy G, et al.: Prevalence and correlates of Chlamydia infection in Canadian street youth. J Adolesc Health 2004, 34:384–390.

24. Noell JW, Ochs LM: Relationship of sexual orientation to substance use, suicidal ideation, suicide attempts, and other factors in a population of homeless adolescents. J Adolesc Health 2001, 29:31–36.

25. Safe Streets Act Revised statutes and consolidated regulations of British Columbia, vol. SBC 75. Canada 2004.

26. Marshall BD, Kerr T, Shoveller JA, Patterson TL, Buxton JA, Wood E: Homelessness and unstable housing associated with an increased risk of HIV and STI transmission among street-involved youth. Health Place 2009, 15:753–760.

27. DeMatteo D, Major C, Block B, Coates R, Fearon M, Goldberg E, King SM, Millson M, O'Shaughnessy M, Read SE: Toronto street youth and HIV/AIDS: prevalence, demographics, and risks. J Adolesc Health 1999, 25:358–366.

28. Greene JM, Ennett ST, Ringwalt CL: Prevalence and correlates of survival sex among runaway and homeless youth. Am J Public Health 1999, 89:1406–1409.

29. Bungay V, Malchy L, Buxton JA, Johnson J, MacPherson D, Rosenfeld T: Life with jib: A snapshot of street youth's use of crystal methamphetamine. Addiction research and theory 2006, 14:235–251.

30. Clatts MC, Goldsamt L, Yi H, Gwadz MV: Homelessness and drug abuse among young men who have sex with men in New York city: a preliminary epidemiological trajectory. J Adolesc 2005, 28:201–214.
31. Small W, Kain S, Laliberté N, Schechter MT, O'Shaughnessy MV, Spittal PM: Incarceration, addiction and harm reduction: inmates experience injecting drugs in prison. Substance use & misuse 2005, 40:831–843.
32. Epperson M, El-Bassel N, Gilbert L, Orellana ER, Chang M: Increased HIV Risk Associated with Criminal Justice Involvement among Men on Methadone. AIDS and behavior 2008, 12:51–57.
33. Zamani S, Kihara M, Gouya MM, Vazirian M, Nassirimanesh B, Ono-Kihara M, Ravari SM, Safaie A, Ichikawa S: High prevalence of HIV infection associated with incarceration among community-based injecting drug users in Tehran, Iran. J Acquir Immune Defic Syndr 2006, 42:342–346.
34. Calverley D: Youth custody and community services in Canada, 2004/2005. Volume 27. Ottawa, Ontario, Canada: Canadian Centre for Justice Statistics, Statistics Canada; 2007.
35. Reuter P: What drug policies cost: estimating government drug policy expenditures. Addiction 2006, 101:315–322.
36. Cunningham JK, Liu LM, Callaghan R: Impact of US and Canadian precursor regulation on methamphetamine purity in the United States. Addiction 2009, 104:441–453.
37. Halkitis PN, Green KA, Mourgues P: Longitudinal investigation of methamphetamine use among gay and bisexual men in New York City: findings from Project BUMPS. J Urban Health 2005, 82:i18–25.
38. Rachlis BS, Wood E, Zhang R, Montaner JS, Kerr T: High rates of homelessness among a cohort of street-involved youth. Health Place 2009, 15:10–17.
39. Ochnio JJ, Patrick D, Ho M, Talling DN, Dobson SR: Past infection with hepatitis A virus among Vancouver street youth, injection drug users and men who have sex with men: implications for vaccination programs. CMAJ 2001, 165:293–297.

Medication Management and Practices in Prison for People with Mental Health Problems: A Qualitative Study

Robert A. Bowen, Anne Rogers and Jennifer Shaw

ABSTRACT

Background

Common mental health problems are prevalent in prison and the quality of prison health care provision for prisoners with mental health problems has been a focus of critical scrutiny. Currently, health policy aims to align and integrate prison health services and practices with those of the National Health Service (NHS). Medication management is a key aspect of treatment for patients with a mental health problem. The medication practices of patients and staff are therefore a key marker of the extent to which the health practices in prison settings equate with those of the NHS. The research reported here considers the influences on medication management during the early stages of custody and the impact it has on prisoners.

Methods

The study employed a qualitative design incorporating semi-structured interviews with 39 prisoners and 71 staff at 4 prisons. Participant observation was carried out in key internal prison locations relevant to the management of vulnerable prisoners to support and inform the interview process. Thematic analysis of the interview data and interpretation of the observational field-notes were undertaken manually. Emergent themes included the impact that delays, changes to or the removal of medication have on prisoners on entry to prison, and the reasons that such events take place.

Results and Discussion

Inmates accounts suggested that psychotropic medication was found a key and valued form of support for people with mental health problems entering custody. Existing regimes of medication and the autonomy to self-medicate established in the community are disrupted and curtailed by the dominant practices and prison routines for the taking of prescribed medication. The continuity of mental health care is undermined by the removal or alteration of existing medication practice and changes on entry to prison which exacerbate prisoners' anxiety and sense of helplessness. Prisoners with a dual diagnosis are likely to be doubly vulnerable because of inconsistencies in substance withdrawal management.

Conclusion

Changes to medication management which accompany entry to prison appear to contribute to poor relationships with prison health staff, disrupts established self-medication practices, discourages patients from taking greater responsibility for their own conditions and detrimentally affects the mental health of many prisoners at a time when they are most vulnerable. Such practices are likely to inhibit the integration and normalisation of mental health management protocols in prison as compared with those operating in the wider community and may hinder progress towards improving the standard of mental health care available to prisoners suffering from mental disorder.

Introduction

Mental health care provision in prisons constitutes an important system of mental health world wide. However, there has been long standing criticism of the care of prisoners with mental health problems and those at risk of self-harm and suicide [1]. Over the last decade a number of organisational and practical changes have been introduced with a view to reforming the system [1,2] with a particular emphasis on the impact of the early stages of custody. Measures which have been advocated and are gradually being implemented include increasing the availability of day care facilities to provide therapeutic settings in which members of community mental health teams (CMHTs) can run appropriate interventions, the expansion of wing-based in-reach services, the engagement of community-based health professionals to assist in promoting continuity of care on entry to prison and post-release, and self care [1,3]. The policy objective behind these changes has been predicated on the notion of equivalence in the range and quality of services available to prisoners and the integration and normalisation with NHS services. Expectations and assumptions behind this new approach include better recognition of the difficulties associated with adjusting to prison life, directing those finding it difficult to cope to appropriate psychological support, greater awareness of and identification of mental health problems, making appropriate referrals, and producing a care management plan (incorporating a medication regime if necessary) for those requiring care.

In spite of the increasing influence of NHS policy and practice, and a willingness to consider the broader determinants of prisoners' health, the notion that prisons can be supportive, healthy environments is at odds with the view that a therapeutic approach to mental health is undermined by an ethos that disempowers and deprives through processes devoted to discipline and control [4]. With estimates that as many as 95% of prisoners have a diagnosable mental health or substance misuse problem or both [2,5], the ability of prisoners to access primary care services and manage a mental health problem represents a basic indicator of the extent to which normalisation of NHS protocols and values may be judged to have been embedded in everyday Prison Service practice. Medication management is a key indicator of the extent to which prison mental health practices equate with those delivered in community settings. Whilst previous qualitative research has considered the factors influencing help-seeking for mental distress by offenders [6], the management and practices of managing medication has not been comprehensively explored. Amongst community populations previous research reports ambivalent attitudes to the taking and prescribing of medication. However, notwithstanding negative side effects, the taking of psychotropic medication for those living in ordinary community settings has been viewed as a key 'prop' in managing mental health. Additionally, shared decision making based on a concordance model which promotes the patients' active involvement has become an adopted norm within mainstream NHS provision [7]. Drawing on the narrative accounts of prisoners and the staff they must negotiate with, this paper considers the prescribing and taking of medication related to the management of mental health problems in a prison context.

Methods

Ethical approval for the study was obtained from the South East Multi-Centre Research Ethics Committee. Data derived from a mixed qualitative methods approach incorporating semi-structured interviews that were supported and informed by participant observation was collected at 4 local prisons in England and Wales during 2004. The establishments comprised a female prison accepting all categories of prisoner (both sentenced and on remand) with facilities for juveniles and young offenders (YOs), a male YO and juvenile facility, a male Category B prison2 and a prison from the High Security Estate accommodating both remand and sentenced adults and YOs. At the time, all were undergoing an evaluated programme of structural and organisational changes intended to improve the management of prisoners believed to be at risk of suicide or self harm.

Table 1. Demographic details of participants

Prisoners		
Gender	Male	27
	Female	12
Age	<25 years	13
	<35 years	17
	<45 years	7
	<55 years	2
Main offence	Drug related	6
	Acquisitive	12
	Violence	14
	Miscellaneous	7
Time in prison	<1 month	12
	<3 months	14
	<6 months	4
	≥ 6 months	9
Experience of F2052SH/ACCT	Yes	32
	No	7
History of mental illness	Yes	29
	No	10
History of self harm	Yes	27
	No	12
Drug/alcohol problem	Yes	25
	No	14
Prison staff		
Gender	Male	43
	Female	28
Role	Chaplain	3
	Detoxification staff	6
	Doctor	3
	Nurses/HCOs	16
	In-reach staff	8
	Social work/out-reach	2
	Prison officer	19
	Probation	1
	Psychiatry	4
	Psychology	1
	Suicide prevention coordinator	7
	Occupational therapist	1

A total of 71 members of staff and 39 prisoners were interviewed [see Table 1]. Members of staff were selected whose daily responsibilities brought them in contact with high-risk categories of prisoner (as described below). These 'key informants' included officers working in reception areas and on induction units, and health care professionals accustomed to managing high-risk patients. A purposive sample of prisoners was selected to provide 'information-rich cases for in-depth study' [8], and to enhance 'situational generalisability' [9]; these included prisoners who:

1. were known to be suffering with or who had a recent history of mental disorder;
2. were currently withdrawing from drug or alcohol misuse;
3. had experience of either the F2052SH4 or ACCT5 processes (or both);
4. had been in prison for at least 2 weeks and less than approximately 8 months.

Semi-Structured Interviews

The interviews lasted approximately 45 minutes to 1 hour, and were recorded on a portable hand-held audio device using micro-cassettes.

Interviews with staff focused on participants' attitudes, and knowledge and training in relation to the identification and management of mental health problems. Staff were asked about their current practices, the division of labour and the impact that the environment had on mental health related work, and were asked about their professional relationships with other members of staff, and with the prisoners that they manage.

Interviews with prisoners explored participants' state of mental health on arrival in prison, their concerns at that time, and how these concerns were met. Prisoners were asked about the environment, regime and practices that they had experienced since entering prison, and the effect that these had had on their mental health. Prisoners were also asked to comment on their relationships with members of staff from various disciplines and their ability to access support networks.

A manual, iterative and reflexive approach to the thematic analysis of the interview data collected during this study involved the repeated review of both the audio recorded interviews and transcribed text to draw out key themes. Tables were then produced to highlight these issues; the tables permitting inter-group (i.e. between establishment/staff grouping e.g. nurses, officers, medical staff) and intra-group (i.e. between individuals within a particular establishment) comparisons to be made, assumptions derived that could be retested in the data collection process, and finally, conclusions drawn. This process closely follows that described by Miles and Huberman [10].

Participant Observation

Structured observations intended to compliment the data accrued from the interviews centred on areas identified as having the greatest impact upon the detection and management of high-risk prisoners in a bid to capture representations of interaction between staff and prisoners. These areas included reception, induction, residential, in-patient and detoxification units. A non-participative approach with the intention of recording as much contextual information (through the noting of verbal comments, descriptions of the physical setting and details of associated processes and events) was adopted. Considerable importance was placed on efforts to merge with and cause minimal impact to the environment being studied. Periods of observation lasted between 2 and 7 hours in each location. Field notes were recorded using pen and paper. Analysis through the interpretation of observed events, consideration of alternative perspectives and meanings, and development of theory was initiated as the observations took place as suggested by May [11]; clarification of recorded events being sought through timely informal query or in the course of interviews with participants. Subsequent to data collection, field-notes were subject to review and re-evaluation with a view to further interpretation and comparison with interview data where possible.

Results

The key themes to emerge from the data included the consequences that disruption to prisoners' medication on entry to prison had on their well-being, the impact of inherent restrictions within prison regimes and practices, illustrations of the ways in which prisoner-patients' autonomy in relation to taking medication is curtailed, the ensuing distancing of relationships between healthcare professionals and prisoners, and problems associated with dealing with comorbidity.

Disruption to Medication Management: A Barrier to Coping with Mental Health and Managing in Prison

Sixteen of the 36 inmates suffering with conditions that included schizophrenia, bi-polar disorder, depression or anxiety expressed grievance about the medication regime that was imposed on entry into prison and the impact that this had on their mental state. The following quotes record the sense of concern experienced by prisoners who, on arriving in prison (some for the first time) were confronted with the realisation that they would have to cope without long-standing medication.

> *"I was on tablets for depression running back over the past 10 years, and when I came here, they refused to give me any...... so for just short of a month of being here, I didn't get any... And when I first came in and I explained it, I explained what medication I was on on the outside, and the doctor says 'well we don't give that out in here.' When he said " we don't give that out in here', I thought 'Whohh! That's what I've always had.....' They were listening but they weren't understanding....... That's how they are in here..... They've got their opinion in their head and nothing's gonna change that." (male prisoner, ID 39)*

The altering of medication without negotiation also created distress as recounted by the following participant who shortly before entering prison had been treated in a hospital psychiatric unit for bi-polar disorder.

> *"I felt I was coping alright with these tablets and then when I knew I wasn't getting any, I just panicked really. The first night I was crying and.. I was beside myself really... because when I was in the hospital, I was on Trazodone (anti-depressant), and they [i.e. specifically the prison doctor] changed it to Venlafaxine. ...and that one I've forgot the name of, for the bi-polar, they just stopped them... It's quite a puzzle to me, 'cos I did get better in there [when previously in hospital], and I can't imagine how I'm going to be alright without it ...." (female prisoner, ID 9)*

Delays in getting access to medication that had been taken on a regular basis (in some cases prescribed during a previous recent stay in prison) were reported to result in a deterioration of individuals' mental states with the ensuing need to incorporate heightened levels of supervision or invoke what was perceived by prisoners to be punitive surveillance as illustrated by the following accounts:

> *"I expected to [i.e. to receive medication], but I didn't take any for......until the end of the weekend..... [for] 4 days...' cos they didn't have any in the pharmacy..... I started going a bit mad, a bit loopy.. [I] self-harmed....And I asked them to put me on 2052, cos I didn't feel well." (male prisoner, ID6)*
>
> *"The doctor told me he wasn't going to give me me anti-depressant ....So I said, all I said was ' it's no wonder people hang their selves.' It was taken the wrong way and I was taken to hospital and put in a 'strip cell' because they thought I'd said that I was going to hang meself.... I tried to explain that I'd only said it out of frustration because I mean, it is a worry. The medication does help. I've tried just about every anti-depressant. I've been on this one for than 3 years now." (male prisoner, ID4)*

Discontinuity between Medication in the Community and Prison: The Importance of Entry Processes

In raising the issue of disruption of prisoner medication on entry to prison, several healthcare professionals who were interviewed cited a number of causes and effects that were felt to contribute to recognised inconsistencies in prescribing practice. Whilst some respondents noted the propensity of some prisoners to lie, a fundamental cause for the lack of continuity in receiving medication on entry to prison was attributed by one in-reach worker to the difficulty of ensuring that new reception prisoners with the greatest needs are seen promptly by the prison doctor. This participant stated:

> *"The only way really around it is that you need to revamp the system of people being reviewed [on arrival in prison]. If you can imagine, the courts sit 'til 5 o'clock. If someone is remanded, they mightn't get to the prison 'til 8 o'clock, 9 o'clock that night. They're [the nursing staff on duty] not going to start ringing GPs at that time of night. In which case, they're then referred to healthcare. If they're lucky, they'll see them the next day. If there's a huge number of people to be seen, they might not be seen for 2 or 3 days. These are where the delays occur."*

The same member of staff also offered an explanation as to why some prisoners' requests to have what they claim has previously been prescribed for them by their community GP (general practitioner) continued in prison, would often be met with a firm refusal:

> *"Where you get the problems is where someone comes in who is clearly going to need a detox also, who immediately starts to tell you that he's been taking Valium and Temazepam, and they've all been prescribed by his GP. You know.. of course they are [sarcasm inferred]. And the number of people that they [i.e. staff] do checks on, and they're not. They've [the prisoner] been buying drugs or whatever. So people tend to be less enthusiastic, shall we say, about making the phone calls and whatever, and just say to people 'I'm sorry, these drugs are just not available in this prison', which is not always correct... Valium is the obvious one. We can use Valium in the prison but it is extremely rare that we use it and it is a 'no-no.' Technically, in here, [it's] a non-formulary item, so you have to fill out another form. You have to get another doctor to agree with you so as to prescribe it, which is time consuming. So 99.9% of the time, they'll just tell you 'it's not available.'.." (member of in-reach team, ID 60)*

The problem of confirming prisoners' claims of having previously received prescribed medication was expanded upon by a nurse:

"...If they come in with drugs that are in their name, have pharmacy labels on them, then they get prescribed you see. But because they don't turn up with any evidence of what they've been taking, it is the problem of checking out with the GP surgeries, who are extremely reluctant I have to say, to give us information of what these guys are taking, so that we can continue that. Unless it was wildly outside the formulary which we adhere to, which is the SSSS formulary [the formulary drawn up by the local Primary Care Trust], we wouldn't be changing it, so there is some protection..." (member of nursing staff, ID 49)

One further medication/treatment related issue that emerged in interviews with health care staff was the chaotic state of paper-based prisoners' medical records. The following comments refer to what was for each establishment, the eagerly anticipated link up with the NHS computer driven patient data system:

"I would say that General Practice in here [in prison] is at about 1980 in terms of comparison with the outside world. The biggest deficit now is the lack of an IT system, an integrated IT system, which means we work entirely off paper notes, and have all the problems of paper notes which are that they are a mess, they are difficult to get information from them quickly... We can't trace back what drugs they've been on without having to trawl through the whole lot. ... Like, all the repeat prescribing has to be hand-written, hand-checked. ... We are really back to where I came into General Practice in 1980. However, we are supposed to have a reasonable computer system up and running by Easter, so hopefully when that all gets on then things like Clinical Governance, chasing through repeat prescriptions, monitoring, will all become a lot easier." (doctor, ID 66)

Curtailing Autonomy and Control Over Self-Medication Practices

The normative routines and practices employed by prisoners to manage their symptoms prior to entering prison were reported to be extremely limited once incarcerated. Prisoners noted the consequences of the perceived lack of flexibility in the prison regime and the limited availability of in-possession medication:

"I only had been taking the Trazodone of a night time [i.e. prior to coming into prison]. I had problems for quite a few weeks [i.e. after entering prison]. I used to get the tablet at 4 o'clock before tea at 5o'clock, and if I took the tablet at 4 [o'clock], by the time I come to 5 [o'clock] I couldn't even get myself off the bed because I was that drugged up on it.... But I've manage to get that moved to 7 o'clock now after a lot of negotiation." (male prisoner, ID 18)

Another prisoner who reported spending the majority of his teenage years and adult life in care or penal institutions, and who had a long history of schizophrenia, contrasted the medication protocols and perceived efficiency of the healthcare service of other establishments with those of the prison in which he was currently residing. When asked if he was feeling the benefit of a change to his medication (the dose having recently been increased by the prison doctor in response to a deterioration in his behaviour which the participant believed had resulted from the prison psychiatrist having inappropriately reduced it on entering prison). The interviewee replied:

"Not at the minute, no, 'cos Healthcare keep messing it up... Well they keep.. not bringing it to me. Not giving it to me... Well we'll see, 'cos I got my medication at 12 o'clock last night...I've been in about 3 weeks and it's happened about 5 times. So we'll have to wait and see what time it comes this afternoon." (male prisoner, ID 20)

The outcome and veracity of this participant's comment was supported by the following observational record:

Shortly after completing the interview with the previous participant i.e. male prisoner ID 20, I was observing the activity on the residential unit just as the afternoon medication round was being completed. It became apparent that the prisoner to whom I had just been speaking had been to see the nurse and had once again found that his medication was unavailable. This resulted in the participant and the wing staff immediately becoming engaged in a heated discussion. The manner in which he was pacing aggressively up and down the landing, and shouting at staff led to the conclusion that there was a strong possibility that he would be reprimanded or restrained for what was clearly angry behaviour borne out of his frustration. I was unable to view the outcome of this tense situation as my escort was ready to leave the house unit before the situation was resolved. Observation 6, Prison X, Friday 4 pm

When medication was received, the lack of personal control over taking it was more curtailed than prisoners were used to, causing disruption to practiced means of controlling their symptoms. Healthcare staff recognised the benefits of providing some in-possession medication but were keen to point out the security and welfare concerns associated with certain drugs being used as currency, and the potential for suicidal prisoners to stockpile supplies. Whilst the suggestion that dispensing medication from the prison pharmacy or on the wings provided useful opportunities for monitoring patients' state of health, this approach was recognised as doing little to develop personal responsibility and was widely recognised as failing to meet many prisoners' needs with respect to the timing that medication should be taken. One doctor noted:

> *"...If I write up a drug [i.e. a prescription for a prisoner] for three times a day, this is one of the issues that we are trying to deal with at the moment, they are going to get 3 doses, some of them, within as little as 8 hours. Whereas again, if you were at home you'd take them breakfast time, lunchtime and evening time, but because of the needs of the discipline staff to be monitoring the queues and things, then our medication regimes have to fit in with them, and it does lead to some friction. We are trying to work on that at the moment." (doctor, ID 66)*

Alienation and Mutual Distrust: Anti-Therapeutic Relationships between Staff and Inmates Over Medication Prescribing

Patient-centredness is a hallmark of high quality primary care within the NHS. In recent years a focus on negotiating medication with patients has become mainstream in primary and secondary care and the notion of a therapeutic alliance over medication and the provision of information has become normative. A majority of the respondents made it clear that they felt there was little point in speaking to the doctor as their requests to have their medication adjusted were generally ignored. One female prisoner who had a history of schizophrenia went further, stating that she was reluctant to engage with medical staff out of fear that complaining might result in her current medication (which had been prescribed by the doctor at the previous prison from where she had recently been transferred) being removed altogether. When asked if she queried the dose that she had been given, she replied:

> *"I didn't bother. I was more concerned about taking my medication, and I didn't want to say anything 'cos I thought if I said anything they might just take me off it... So I just kept my mouth shut basically. I daren't say anything. You know what they're like." (female prisoner, ID 17)*

The perceived arbitrariness of prescribing practice was central to the frustration and heightened anxiety experienced by many prisoners, and was identified as contributing to the strained relationships between inmates and healthcare professionals - a point illustrated by a participant who remarked:

> *"Yeah.. with prison and the 'out' [outside community], it's different. Like, on the out, your doctor knows who you are, what you are, what medication you're on and what your problem is. In here, it doesn't matter what medication you're on*

out there, you don't get it in here. Do you know what I mean?" (male prisoner, ID 20)

Whilst this participant's comment makes indirect reference to the benefits that follow from there being a history of contact with one's GP - an association that rarely develops during comparatively short stays in local prisons, one nurse alluded to the comparative ease with which difficult and potentially unpopular decisions regarding patients' treatment regimes could be implemented in prison settings, stating:

"The standard of care [medical care] is good, and I would think that some of the inmates would think it was good, but a lot of them would think it was bad because they're not getting what they get on the 'out.'.. If a doctor is in his surgery on a little estate somewhere and someone comes in screaming and shouting for something, and he feels intimidated and wants his surgery to be nice and quiet, he'll give them a script, a prescription, and he's got them out the door.... But if you're in a place like a prison, where they can't go anywhere, they can't be disruptive or if they are disruptive, they can be removed, then you can say 'no, I'm not going to give you that drug.' And so I think that the general consensus might be that we've got rubbish doctors because 'the doctor on the 'out' would give me it.' But it doesn't necessarily mean that the doctor on the 'out' is good, it's not his fault but a lot of people get pacified on the 'out.' People get kept on Valium for years and it shouldn't happen." (nurse, ID 8)

Despite the level of concern expressed by prisoner participants in relation to their medication being changed, reduced or stopped completely, few health care staff made direct reference to these issues, and fewer still commented on the effects that such occurrences might cause. However, a prevalent theme among those that did contribute to this issue highlighted their concerns that prisoners were often disingenuous in their claims that they had been receiving some prescription drugs, as illustrated by the following quote:

"Like there's one guy at the moment who is convinced that he's on certain doses of certain things and I've got the GP to read me his psychiatrist's letter that came in January, so I know that the doses we've prescribed are correct. Do you know what I mean? 'Cos I've seen him three times with the same issue... So there's a bit of that, and a bit of manipulation..." (member of nursing staff, ID 49)

The suspicion contained within the previous participant's dialogue was supported in the frank opinion of a psychiatrist:

"I think the big difference between civilian psychiatric practice and working here is that in civilian psychiatric practice people rarely actually lie to you. I mean, they highlight things they want you to be aware of and minimise things they don't want you to be aware of. I suppose it's lying really, but usually there's a kernel of truth in 95% of cases; whereas in here, 95% of the people that you're speaking to are telling you things that aren't true. That's a politically incorrectly explanation but ... The aim usually is to obtain either pain killers or opiates such as Co-codamol, just to get some kind of sedative so that they can basically blot out reality really... It's quite crucially important really [to understand what is going on] 'cos what happens is that if the doctors who are involved just give in when they [the prisoners] come in and start ranting and raving about opiates and so on, and the doctor kind of goes 'okay' and gives in to them then it makes it harder for the prison staff 'cos he goes back and tells the wing that Dr X is a walkover and then they are all coming over, and if they get codeine out of the doctor, they sell it for 'gear' [i.e. drugs] to other prisoners and it makes a breakdown of the system more likely." (psychiatrist, ID 26)

Evidence of the dishonesty employed by prisoners and the potential for confrontation were recorded in the following extract of an observation of a health screening interview conducted with a newly arrived prisoner on the reception unit of one of the prisons:

'...Throughout the interview, the prisoner appeared upset, avoided eye contact (looking down at the table much of the time), telling the nurse that s/he was 'rattling' i.e. suffering from the withdrawal from drugs, and couldn't be bothered answering the nurse's questions, other than to say that s/he had previously seen a psychiatrist but was not willing to say what for, or when. The prisoner then asked for his/her own medication that s/he had brought from court. This was refused, the nurse explaining that s/he would need to be seen by the doctor first. This was met by abuse - the prisoner shouting that s/he wanted them now... Approximately 10 minutes later, the prisoner was observed sat at a desk in reception talking to an officer who was noting personal details. The prisoner was seen to be relaxed, chatting calmly and joking with an officer.

Two hours later, the prisoner, returned to the reception unit to see the doctor as s/he came on duty. In the interview room, the prisoner immediately re-adopted the demeanour s/he had demonstrated when s/he had seen the nurse. Clutching a handkerchief to his/her mouth, his/her hands and legs were seen to shake. The prisoner avoided eye contact and responded to questions by a nod or shake of the head. The doctor focused on his/her current medication (which was recovered from her possessions) and the prisoner's drug problem. The doctor queried the prisoner's use of Amitriptyline, and s/he admitted that this was not for depression but more to help him/her sleep. The doctor then explained how a Methadone detox would be given but that it would start at 10 mg, rising to 30 mg and then tailing off. At this point the prisoner was very quiet. The doctor also explained that Diazepam would be given but it would be administered in 15 mg doses, one in the morning and one at night. Realising that the total quantity was much less that s/he had been used to, the prisoner remonstrated with the doctor. The prisoner was also told that although s/he would be given Amitriptyline that night, it would be reviewed by the doctor the following day. At this point, the prisoner became more agitated. The interview ended with the prisoner being led away, clearly disgruntled.'

Observation 5 - Prison Y, approximately 16.30 hrs

Dealing with Co-Morbidity and Managing Withdrawal from Drugs

Twenty-five of the prisoners interviewed recorded a past history of drug or alcohol abuse. Although not everyone entering prison either requires or requests assistance with withdrawal, those who test positive to having an ongoing substance misuse problem are invariably offered a chemical detox. Participants' comments were found to focus on several areas of concern. Most significant was the variation in practice adopted by different prisons. The following prisoner's comments relating to the prescription of pain relief during detoxification typify the experience of many others:

"I only started getting them 3 days after I came in. I had to wait for my medical records from GGGG [name of prison from where the prisoner had just been transferred] and until that came they couldn't give us any medication. Thing is, I'd been on Methadone there...Yeah, it's different in different gaols.... Like in GGGG, if you're on a script on the 'out', they give you what they call a 'maintenance script' inside, of a smaller dose. Whereas I was on 50 ml on the outside, so in GGGG I was getting 30 ml of methadone and a sleeping tablet. And that was it. That was doing it. But when I came here, they told me they don't do Methadone ..., they don't give you sleeping pills. It's a total no-no. So I was ill, very ill." (female prisoner, ID 15)

Being moved from one prison to another, either for permanent transfer or in order to facilitate an appearance at court in relation to offences committed elsewhere around the country was once again found to cause disruption to prisoners' medication. One such example emerged in the comments of a participant who had been started on a course of pain relief to help with his detox from heroin when he first arrived in prison. He stated:

> "*... but when they shipped me from here to PPPP [a prison nearer to court], my detox medication, I never got that for three days.*"

The same participant went on to recall how during the 3 weeks he was at the other prison, he received medication to try to stabilise his mental condition. However, by the time he returned to the original establishment, although he had completed his detox, his mental condition deteriorated, causing him to be placed on the prison's in-patient unit for 2 days. The following comment once again highlights the lack of support felt by the interviewee from residential unit officers at a time when he was evidently feeling unstable, and highlights the number of different residential wings to which he had been assigned during the first 6 - 8 weeks that he had been in custody. He recalled:

> "*... I approached a member of staff and told him I was still feeling a bit dodgy, mentally. And they didn't want to deal with the issues, they just shipped me straight off to another wing. And then I went, approached the staff on the other wing, told them, and they kicked me off that wing, they didn't want to deal with it, put me back on five [residential unit 5]. And then five put me on here [residential unit 3].*" *(male prisoner, ID 37)*

One other area of complaint to emerge in prisoners' dialogue relates to the absence of information provided by health care staff concerning the medication that they were given. Some prisoners clearly felt they were being patronised and their legitimate interest in the drugs they were being given was being disregarded by some health care staff. Whilst this approach on the part of some health care professionals may result from a wish to avoid confrontation or genuinely result from the view that the treatment being suggested is perceived to be in the patient's best interest, it belies the extensive pharmacopoeic knowledge that many prisoners who have struggled with enduring mental illness or substance misuse are likely to have, and demonstrates not only a lack of appreciation for the need for effective communication but appears to depart significantly from what would be considered good practice in the wider community. One prisoner described a conversation that took place during the health-screen interview. He recalled:

> "*...this time when I came in I said 'I've got a heroin habit.' [The doctor] Said 'right, you'll be doing a Subutex detox.' 'Fine.' And I said to the doctor 'I've been on the crack as well.' And he went 'Right, well take five of these green pills.' I said 'what are they? I like to know what they are.' He said 'they're happy pills, they'll*

make you feel better.' And that was it, and I was told to go into the waiting room again." (male prisoner, ID 26)

A major concern of staff involved in the day-to-day management of prisoners withdrawing from substances misuse echoed that of prisoners in so far as inconsistency in prescribed treatment regimes was associated with increased confrontation. One Health Care Officer (HCO - prison officer who has undergone some level of nursing training) stated:

"I think it's a good thing as well that we're getting a dedicated detox unit. Er.. the thing is with the doctors we've used, the [GP] practice we had before and the [GP] practice we've got now, the detox is too erratic. You know, one doctor will give the 9-day detox, another will cut it down to 3 days or something... It doesn't work...I mean, it's very confrontational at the treatment room a lot of the time.... And there was like transfers [from another prison] in yesterday. ... And they all came up on Methadone detoxs. And we don't use Methadone here so consequently they went from a standard Methadone detox of 25 mg a day to the year dot [sarcasm inferred], to a 9-day liquid DF detox here and then that's it. So they were all creating hell last night when they came in." (HCO7, ID 32)

Equally important however, from a staff perspective, was the need to establish a practicable protocol for dealing with dual-diagnosis clients; the comments of in-reach and detox. staff indicating how creating a clear understanding of the division of labour and developing effective lines of communication (cornerstones of inter-disciplinary working) were essential in ensuring that available resources were appropriately tasked to support patients appropriately. The following quote describes how newly arrived prisoners, recognised by reception health staff to have a mental health and substance abuse issues, might be referred to various support services:

"Well they tend to refer to.. like if someone has drug problems, mental health problems and self-harm, they will refer to us [detox], to CMHT [Community Mental Health Team], and to Out-reach [social support team]. Then the three of us have the referral and we all tend to see them the next morning, and we then try to come to some sort of plan together. That's the way it should work."

However, when queried if the three teams come together in some form of case conference, the interviewee replied:

"No not really. It's difficult. We tend to see them quite quickly whereas CMHT, unless it's urgent have a 3 day [waiting list]. So we often see them first and it depends on the drug use, because there may be drug use and mental health problems

but the drug use might just be that they smoke cannabis at the weekend and have a severe mental illness, and those cases would be taken over primarily by CMHT but those with a very heavy drug use, we take on and then do the mental health assessment. And then we liaise with them and make them aware, and if we have concerns then we will speak to the psychiatrist ourselves. So it depends on the individual and their risk and their needs really but sometimes we're all working with the same person..."(detox staff, ID 52)

Discussion

The health behaviours, clinical and demographic background of prisoners make an important contribution to health in prisons. However, the environment, the regime, and the organisational culture are likely to be more important [1,2,12]. Medication practices are a key indicator of and contributor to the therapeutic prison environment. They have particular relevance in light of findings that approximately 20% of male and 50% of female prisoners take some form of psychotropic medication [13], and that the taking of mood-modifying medicines such as minor tranquillisers and other psychotropic medication provides support and encourages patient engagement. An important element of the later is to encourage patients to participate in daily decisions about treatment which in turn is perceived to be a key part of their recovery [7].

This study found a tension in the standards and nature of official policy concerning mental health and what is happening on the ground. Prison policy espouses the goal of delivering mental health services to prisoners that provide 'effective through-care that responds quickly and seamlessly to their changing needs' [2], yet a common theme to emerge in the descriptions of both inmates and prison staff indicated discontinuity in medication treatment received on entering prison. Such findings are contrary to the purported goal of seeking to normalise mental health care in prisons to equate with the norms, values and practices existing in the wider community. Delays, stoppages or changes to medication were noted to be the underlying causes of confusion, anxiety and distress reported by half of prisoners interviewed. At a time when prisoners are perceived to be in a particularly vulnerable state and experience the loss of normal social support, such actions should be recognised as representing the removal of a prop. Restrictions to self-medicate further limit individuals' opportunities to engage in self-medication and management that are generally available to them in community settings.

The role of staff in providing high quality mental health care has been highlighted by HM Chief Inspector of Prisons review [5]. This report stresses the importance of relationships between prison staff and prisoners, and warns of the danger of staff failing to recognise the impact that entry to prison for the first time can have. Notwithstanding the deficits associated with local prisons (the high turnover of the prison population, overcrowding, inadequate resources to provide purposeful activity etc) and the difficulties encountered by nursing and medical staff working in an environment in which 'healthcare culture is influenced by traditional attitudes, with an emphasis on security [1], the importance of doctor-patient communication has been suggested as playing an important role in securing effective treatment outcomes. Indeed, it has been suggested that patients have multiple needs which fail to be voiced due to doctors misinterpreting what their patient's agendas actually are [14]. The findings of this study confirm this to be particularly so in prison settings where pre-conceived notions of prisoners' objectives, the limited amount of time available to conduct patient interviews, and doctors' attempts to 'fit in' with established prison practice [1]. The latter often resulted in frequently rushed, impersonal consultations in which little attention appears to be afforded to prisoners' concerns and little interest shown in discussing treatment options. The dysfunctional nature of such interactions both promote and perpetuate prisoners' views of health care as being another form of discipline, whilst they themselves are generally perceived as 'problematic' or 'malingerers' [15].

Conclusion

One risk that requires managing in prison settings is the obtaining and use of prescription drugs ....for which there is no medical justification. Accounts of staff and observations carried out in the course of the present study suggested that a proportion of prisoners may attempt to deceive healthcare staff in order to obtain prescriptions. An unfortunate corollary of this is that prisoners as a group are typecast as being untrustworthy and manipulative; such labelling consigning those who present with genuine mental health problems (as indicated by formal screening and previous community management) to greater suffering, loss of control, deterioration in mental health state and risk. Whilst some ....prison healthcare staff have a role to play in deciding on prescription changes or encouraging self medication practices which are conducive to patients' routines and needs, it is likely that even those performing to the best of their ability are likely to be constrained by organisational, environmental and cultural factors which currently restrict or pose barriers to introducing standards of service that are commonplace in the wider community and to which the Prison Service aspires. The distribution of in-possession medication and control over self-medication is to be promoted to those who would benefit. This is already accepted working practice in some establishments; risk assessment and the establishment of appropriate protocols having been incorporated into working practices. The supply of prescription drugs which may be subject to subsequent misuse remains a risk for prison staff that needs to be managed. However, findings suggest that attention needs to be given to removing institutional barriers, and changing professional practices and interactions with inmates with mental health problems in a way that is more therapeutically orientated. The limited opportunity for consultation during rushed reception procedures, the availability of appropriately trained and experienced medical staff at such times, restrictions to patient contact due to the prison regime [16], and the technological improvements to modernise prison records and IT systems are just a few examples that undermine the effective care of prisoners. In the absence of the organisational changes required to effect more flexible working practices [17] and afford healthcare staff the time they need to 'unpack' prisoners' health matters and engage in dialogue that addresses their concerns, this paternalistic approach to restricting prescribed medication will continue to the detriment of those in greatest need and hamper progress towards the goal of ensuring that prisoners receive an equivalent standard of care to that more widely available in the NHS.

Despite evidence of the scale of mental disorder among prisoners and an acute awareness of long established deficits in the standard of prison healthcare, much of the progress towards addressing the inherent vulnerability of such prisoners appears to have been slow in its implementation limited. Current moves to introduce mental health in-reach teams and dedicated detox. facilities are an important step to improve patient access to specialist services. Nonetheless, their effectiveness is likely to be undermined if there is no change in routine practices and a greater awareness of the need and opportunities for prisoners to take personal responsibility for their treatment. The latter is relevant for developing therapeutic relationships and the nurturing of ideologies that support the interests of a more therapeutic regime [18].

Competing Interests

The authors declare that they have no competing interests.

Authors' Contributions

RB designed and modified the qualitative study and undertook the field work and drafted the manuscript, AR designed the study, assisted with data analysis and interpretation and drafted the manuscript, JS designed the larger study within which this project was based and contributed comments to the manuscript.

Acknowledgements

We would like to acknowledge funding for undertaking the project from the Home Office and NIHR. The views expressed are those of the authors. We are grateful to respondents and staff in each of the prison settings for agreeing to be interviewed and giving of their time.

References

1. HMPS/DoH: The Future Organisation of Prison Health Care. London: Department of Health; 1999.
2. HMPS/DoH: Changing the Outlook: A Strategy for Developing and Modernising Mental Health Services in Prisons. London: Department of Health; 2001.
3. Gately C, Kennedy A, Bowen A, Macdonald W, Rogers A: Prisoner perspectives on managing long term conditions: a qualitative study. International Journal of Prisoner Health 2 2006, 2:91–99.
4. Hughes RA: Health, place and British prisons. Health & Place 2000, 6(1):57–62.
5. HMIP: Suicide is Everyone's Concern: A Thematic Review by HM Chief Inspectorate of Prisons for England and Wales. London: Home Office; 1999.
6. Howerton A, Byng R, Campbell J, Hess D, Owens C, Aitken P: Understanding help seeking behaviour among male offenders: qualitative interview study. British Medical Journal 2007, 334(7588):303–306B.
7. Rogers A, Pilgrim D, Brennan S, Sulaiman I, Watson G, Chew-Graham C: Prescribing benzodiazepines in general practice: a new view of an old problem. Health 11 2007, 2:181–198.
8. Patton MQ: Qualitative evaluation and research methods. Newbury Park, CA: Sage; 1990.
9. Hornsburgh D: Evaluation of Qualitative Research. Journal of Clinical Nursing 2003, 12:307–312.
10. Miles MB, Huberman AM: Drawing Valid Meaning from Qualitative Data: Toward a Shared Craft. Educational Researcher 1984, 13(5):20–30.
11. May T, Ed: Qualitative Research in Action. London: Sage; 2002.
12. HMPS: Report of the Director of Health Care, 1995–1996. London: The Stationery Office; 1997.

13. Singleton N, Meltzer H, Gatward R: Psychiatric morbidity among prisoners in England and Wales. London: Office for National Statistics; 1998.
14. Barry CA, Bradley CP, Britten N, Stevenson FA, Barber N: Patients' unvoiced agendas in general practice consultations: qualitative study. BMJ 2000, 320:1246–1250.
15. HSJ: Good Management. [http://www.hsj.co.uk/healthservicejournal/pages/GM2/p26/070614], 2007.
16. Norman A, Parrish A: Prison health care: work environment and the nursing role. British Journal of Nursing 1999, 8(10):653–656.
17. Nurse J, Woodcock P, Ormsby J: Influence of environmental factors on mental health within prisons: focus group study. BMJ 2003, 327:7413–480.
18. Gesler W: Therapeutic Landscapes: Medical Issues in Light of the New Cultural Geography. Social Science & Medicine 1992, 34(7):735–746.

Estimating Incidence Trends in Regular Heroin Use in 26 Regions of Switzerland Using Methadone Treatment Data

Carlos Nordt, Karin Landolt and Rudolf Stohler

ABSTRACT

Background

Regional incidence trends in regular heroin use are important for assessing the effectiveness of drug policies and for forecasting potential future epidemics.

Methods

To estimate incidence trends we applied both the more traditional Reporting Delay Adjustment (RDA) method as well as the new and less data demanding General Inclusion Function (GIF) method. The latter describes the probability of an individual being in substitution treatment depending on time since the onset of heroin use. Data on year of birth, age at first regular heroin use and date of admission to and cessation of substitution treatment was available from 1997 to 2006 for 11 of the 26 regions (cantons) of Switzerland. For the remaining cantons, we used the number of patients in 5-year age group categories published in annual statistics between 1999 and 2006.

Results

Application of the RDA and GIF methods on data from the whole of Switzerland produced equivalent incidence trends. The GIF method revealed similar incidence trends in all of the Swiss cantons. Imputing a constant age of onset of 21 years resulted in almost equal trends to those obtained when real age of onset was used. The cantonal incidence estimates revealed that in the mid 80s there were high incidence rates in various regions distributed throughout all of the linguistic areas in Switzerland. During the following years these regional differences disappeared and the incidence of regular heroin use stabilized at a low level throughout the country.

Conclusion

It has been demonstrated that even with incomplete data the GIF method allows to calculate accurate regional incidence trends.

Background

A lack of information about trends in the incidence of regular heroin use hinders effective drug policy and public health action. Yet incidence trends are still unknown almost everywhere in the world. Various estimation methods have been developed to estimate the incidence of heroin use [1-4], however they have requirements, like long-term treatment data or reliable drug mortality statistics, which are rarely available. We have recently reported estimates of incidence and prevalence trends in regular heroin use in Zurich, Switzerland [5]. These estimates were produced using the Reporting Delay Adjustment (RDA) method and were based on data from a long-term case register that covered all methadone treatments for more than a decade. The RDA method led to prevalence estimates that were in good accord with prevalence estimates generated by other approaches. We have also developed a simpler procedure to estimate the incidence of regular heroin use, called the General Inclusion Function (GIF) method, which only requires methadone treatment data from a single day [6]. On the premise that heroin dependence is usually a chronic condition, we have hypothesized that if governmental regulations do not restrict access the probability of an individual being in substitution treatment depends largely upon time since onset of regular heroin use. The GIF approach led to reasonably good incidence estimates in the canton of Zurich, despite the presence of open drug scenes and irrespective of whether onset of regular heroin use occurred during an early or a late phase of the 'heroin epidemic.'

To explore if the GIF method yields plausible incidence estimates in other regions, we decided to apply it to other Swiss cantons. There are several advantages of using other areas in Switzerland to test the GIF method. First, even though incidence trends in regular heroin use are unknown in almost all regions, drug mortality trends and the stable number of patients in substitution treatment suggest that prevalence trends are similar throughout Switzerland. Second, each of the 26 cantons in Switzerland is individually responsible for treatment provision and data collection. Third, the size of the population of regular heroin users differs vastly between cantons. For example, in 2003, the annual number of individuals in methadone treatment in various cantons ranged from 3 to 3,592 [7]. Therefore, by using separate datasets that should yield similar incidence trends in each canton, we can determine the smallest area to which the GIF method can be applied.

Unfortunately, not every Swiss canton has a methadone treatment register that includes data on the year of first regular heroin use. Since heroin use usually occurs early in life [1], when data on the age of heroin onset are missing, an estimate of the number of affected individuals in each birth cohort may be used as an approximation for incidence estimates. A benefit of using data collected on the year of birth of patients in substitution treatment is that it probably contains fewer errors than age at first regular heroin use. If we find similar incidence estimates in several adjacent regions and similarly affected birth cohorts, estimates on birth cohorts in the remaining regions would be sufficient to draw a rough picture on how a 'heroin wave' has spread over the whole area.

Thus, the aims of this paper are to: (i) estimate the incidence of regular heroin use for the whole of Switzerland using the Reporting Delay Adjustment (RDA) method; (ii) ascertain a General Inclusion Function (GIF) for the whole of Switzerland by using the annual number of patients in substitution treatment; (iii) apply the GIF function to treatment data from different calendar years, using data on 'year of first heroin use' as well as 'year of birth'; and (iv) examine if assuming a constant age of onset leads to acceptable incidence trend approximations in all Swiss cantons.

Methods

Study Area

Open drug scenes started to develop in Switzerland in the early 1980s, mainly in the city of Zurich, but also in other German speaking parts of Switzerland. The

French and Italian speaking parts of Switzerland were not affected. Thus, linguistic regions may differ concerning the development of the heroin wave. In February 1995, the last open drug scenes in Zurich, Solothurn, and Olten were closed; those in Bern, Basel-Stadt, and St. Gallen had been closed earlier [8].

Methadone Treatment Registers

The Swiss law on narcotics requires that treatment providers obtain permission from cantonal health authorities to prescribe opioids to people dependent on heroin. The law also requires a register of substitution treatments provided to opioid dependent persons. Each of the 26 Swiss cantons is responsible for methadone treatment provision and data collection. The Swiss Federal Office of Public Health has collated and published annual statistics from all cantons since 1999 [7]. Since 2002, it has provided a computer program to all interested cantons by which more detailed treatment data can be collected and subsequently accessed. We obtained most of the register data via the Swiss Federal Office of Public Health. Variables included a personal identification number, date of birth, age at first regular heroin use, date of admission to and cessation of each treatment, and date of the first lifetime treatment episode. Although the same database was used, most, but not all, cantons supplied time-ranges of permissions to treat a specific patient, lasting several months to a year. Moreover, cantons collected age at first regular heroin use to varying degrees.

Data from this database and other datasets were analyzed, if cantonal health authorities agreed to participate in the study. The following 11 cantons were enrolled: Bern, Fribourg, Genève, Graubünden, Neuchâtel, Schaffhausen, Solothurn, St. Gallen, Ticino, Vaud, Zug, and Zurich. Additionally, we obtained the exact year of birth of all patients in treatment on a specific day for 2 cantons (Aargau, 28th of January 2008, n = 812; Solothurn, 17th of January 2008, n = 621). We did not approach most of the remaining 13 cantons because they had limited sample sizes, generally having less than 100 individuals in treatment. Data from the 11 cantons fully included in this part of the study represented about 70% of the Swiss population in methadone treatment. These 11 cantons were spread throughout different linguistic areas and included both urban and rural parts of Switzerland.

Data Preparation

We examined the data for consistency. If heroin onset was recorded to be before the age of 12, if onset was not recorded to be in the year of first substitution treatment or before, or if discrepancies between a patient's year of onset recorded in different treatment episodes were greater than 3 years, the year of onset from this treatment episode was considered as an error in the data and thus deleted. If there were discrepancies in the year of onset for an individual patient of 3 years or less, a mean value was calculated and rounded accordingly. For the remaining patients, we assumed that the missing data did not strongly deviate from the data obtained for the 14,396 (53%) patients with a known year of first regular heroin use. For the RDA method we used all available data from the 11 cantons included in this study. For the GIF method we only used data from those calendar years that had similar numbers of admissions and cessations. This ensured that in the analyses, we only included data from years in which a register had been fully operational.

Calculating Incidence Estimates using the RDA Method

The RDA method aims to estimate and to correct for those heroin users from each onset cohort, who have not yet shown up in opioid substitution treatment. It assumes a stable probability distribution that describes the time between onset of heroin use and first treatment. If we assume that almost every regular heroin user enters his first substitution treatment (if ever) within 10 years, a treatment register of ten years would be sufficient to estimate a full probability distribution of that 'delay' or 'lag.' To estimate the lag time period between first regular heroin use and first substitution treatment we applied the method proposed by Hickman and colleagues [2]. To avoid bias due to restricted availability of treatment slots or missing data before 1997, we only included individuals (n = 2,709), who had begun using heroin regularly between 1997 and 2006. We estimated the conditional lag distribution of this 10-year period with a general linear model using five parameters (a linear and a quadratic term of the lag time and dummy variables for the three shortest lag time periods), as in a former publication [5].

The RDA method underestimates incidence because it does not account for those heroin users, who do not show up in treatment. Thus, we applied a cessation correction of 4% per year, based on the assumption that each heroin user observed in treatment is representative of all those who had died or stopped heroin use before having had the possibility to enter treatment. For example, if a total of 44 individuals, who entered treatment in 1997, reported initiating regular heroin use in 1977, this suggests that 100 individuals had started regularly using heroin in 1977 (= $0.96^{20\ \text{years}}$).

Estimating the Annual Probability of Being in Substitution Treatment (GIF)

We divided the annual number of individuals in substitution treatment by the RDA incidence estimate for each calendar year between 2001 and 2006 (n =

12,846 to n = 13,336). Since the data revealed a preference for reporting even years of regular heroin onset, we computed a three-period moving average using the former, the current, and the following values. The resulting treatment probabilities were plotted by time since onset of regular heroin use.

Applying the GIF to Year of First Regular Heroin Use

To estimate incidence for each calendar year separately, we applied an approximation formula for the General Inclusion Function. This formula describes the probability of being in treatment at least 1 day during a year (P_{annum}), depending on time since onset (in years):

$$p_{\text{annum}} = 0.7 * (1 - (\text{time} - 0.2)^{-1}) * 0.96^{(\text{time}-0.2)} \,|\, \text{time} \geq 2 \text{ years}.$$

The components of this approximation formula can be interpreted as follows: 0.7 is a linear scale factor and sets the peak to 0.457; the term $(1-(\text{time}-0.2)^{-1})$ describes the steep rise in probability during the first years; after some decades the decline approximates 4% per year ($0.96^{(\text{time}-0.2)}$).

Estimating the Number of Affected Individuals in Each Birth Cohort

The mean age at first regular heroin use was approximately 21 years in all cantons over all treatment years. Thus, we estimated the number of affected individuals in each birth cohort in cantons providing detailed data by supposing a constant age of onset of 21 years and applying the approximation formula for the General Inclusion Function.

Approximating Incidence, Using Estimates of the Number of Affected Individuals in Each Birth Cohort

Shifting the time scale by 21 years–the mean age of onset of regular heroin use–allowed comparisons to be made between incidence estimates based on age of onset data and those based on birth cohort data. Theoretically, the narrower the age of onset distribution, the more similar the two forms of incidence estimates.

Estimating the Number of Affected Individuals in Each Birth Cohort, Using Data in Age Groups, Published in Annual Treatment Statistics

Between 1999 and 2006, 24 out of 26 cantons published the number of patients in treatment by 5-year age group categories over at least several years [7]. To estimate how many individuals of a specific age were in treatment per year, we divided the published number of patients within each age group by 5. We assumed that the resulting number was the best estimate for the size of the specific group, whose age corresponds to the mean age of the category. We obtained a value for the number of patients aged 22, 27, 32 and 37 for each treatment year between 1999 and 2006. This procedure led to the number of individuals in treatment for each birth cohort between 1962 and 1984. When we obtained more than one number for a specific birth cohort, we used the value from the most recent treatment year. Although only 7 out of 24 cantons had published data on age groups for each year between 1999 and 2006, 15 out of 24 cantons had data available from 2002 to 2005. For the canton of Basel-Stadt we had to rely on a publication that only provided 10-year age group categories [9]. Therefore, for this canton we modified our procedure accordingly.

In cantons for which we could recalculate the numbers of patients in treatment using treatment case register data, we found some differences in the size of age groups. Since we were interested in how well the procedure using 5-year age group categories fitted estimates using numbers of patient of each birth cohort from treatment case registers, we used our own calculated annual numbers of 5-year age group categories.

Results

For 11 cantons we had complete methadone treatment register data from 1998 to 2006 (Table 1). The mean annual number of patients treated during the respective years varied from 3,701 in the canton of Zurich to 79 in the canton of Zug. The mean age was approximately 35 years and the mean age of onset of regular heroin use was approximately 21 years. The proportion of cases with known age of onset ranged from 6% to 90% between cantons. In most cantons and during all treatment years the mean age of patients with unknown age of onset was about 3 years higher than the age of patients with known age of onset.

Table 1. Methadone treatment case-register data.

	Treatment years included	Number of patients (mean)	Age (mean)	Age at first regular heroin use (mean)	Proportion with known onset (mean)
Zurich	1998–2006	3,701	35.4	21.5	76%
Bern	1999–2005	2,650	34.8	20.9	50%
Genève	1999–2006	1,581	37.3	20.1	16%
Vaud	2001–2006	1,684	36.0	20.8	36%
Ticino	1998–2006	987	35.5	20.8	90%
St. Gallen	2001–2007	910	35.2	22.9	31%
Neuchâtel	2000–2006	689	36.1	21.1	32%
Fribourg	1999–2005	496	33.4	20.2	51%
Graubünden	1998–2006	267	34.3	19.7	6%
Schaffhausen	2000–2006	140	33.9	19.2	28%
Zug	2006	79	36.2	20.8	72%

Methadone treatment case-registers data from 11 cantons of Switzerland, 1998–2008.

Data from the case registers of all 11 participating cantons showed that in total 27,047 patients had entered substitution treatment until the end of 2006. The RDA method revealed that heroin users entered their first substitution treatment soon after onset, i.e. 22.6% entered two years after onset (Figure 1), if they did so within the period of 10 years that was covered by the data. Thus, summing the lag-time distribution up to year two, we see that one half of heroin users (49.5%) entered substitution treatment for the first time within two years of onset.

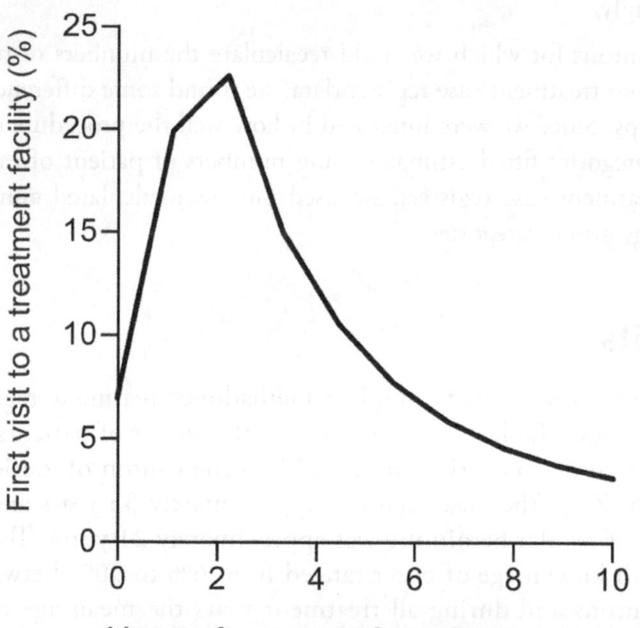

Figure 1. Lag-time distribution (RDA). Joint lag-time distribution between onset of regular heroin use and first visit to a substitution treatment facility in the 11 cantons of Switzerland that provided year of onset, 1997–2006.

Adjustment of the observed incidence number by the lag-time distribution and cessation correction only affected the overall shape of the heroin incidence curve to a small extent (Figure 2). The adjusted incidence curve indicates that heroin use first began in 1966 with 15 individuals. Following this there was a steep rise in heroin incidence, which peaked in 1990 with 2,572 new users, and then a steep decline to 686 users in 2002. Incidence levels have remained stable since then.

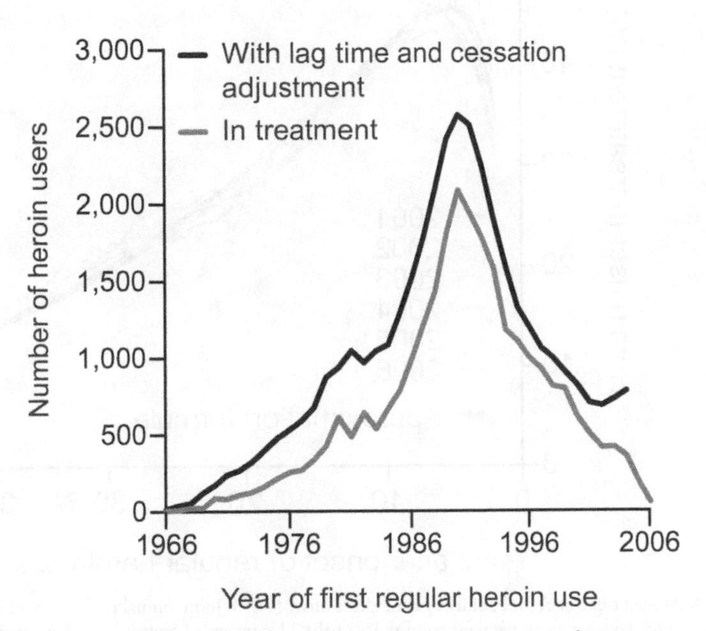

Figure 2. Incidence by the RDA method. Joint incidence of regular heroin use in the 11 cantons of Switzerland that provided year of onset estimated by the RDA method, 1966–2006.

Applying the GIF approach, a steep increase in the proportion of regular heroin users in treatment is observed during the first years after onset (Figure 3). Between 2001 and 2006, 5 years after onset of regular heroin use between 42.8% and 50.2% of users were in substitution treatment for at least 1 day. The proportion of heroin users in treatment declined slowly to about 21% three decades after onset.

The left panel in figure 4 displays the resulting incidence estimates when the GIF approximation formula is applied to the joint dataset of the 11 cantons. It shows that this formula yields similar results using the annual number of patients from 2001 to 2006, respectively. The right panel in figure 4 shows the results of the application of the approximation formula on year of birth when imputing a constant age of onset of 21 years instead of real age of onset. Again, the approximation formula yields similar results using the annual number of patients from 2001 to 2006, respectively. Results indicate that with approximately 2,000 individuals in each cohort, the birth cohorts between 1965 and 1969 were most affected by regular heroin use.

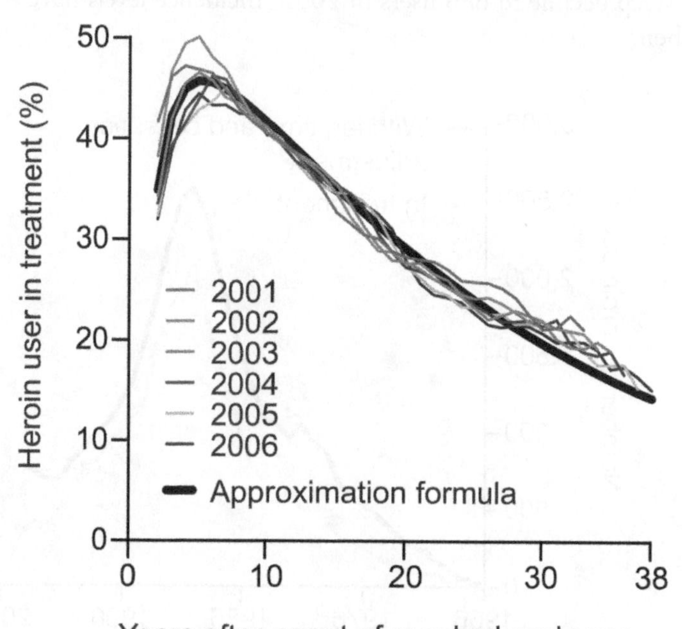

Figure 3. Annual treatment probability by time since onset (GIF). Joint annual probability of being in treatment for at least one day among all regular heroin users of the 11 cantons of Switzerland that provided year of onset, 2001–2006.

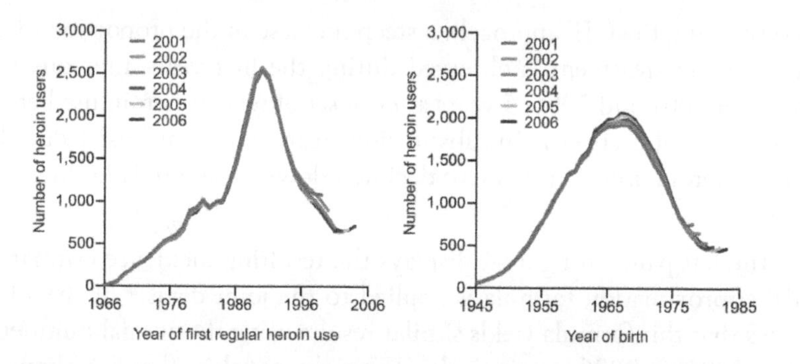

Figure 4. Incidence and affected birth cohorts by the GIF method. Joint incidence of regular heroin use and affected birth cohorts in the 11 cantons of Switzerland that provided year of onset; annual estimates by the GIF method using number of patients, 2001–2006.

The annual incidence estimates from all 11 cantons were very similar, especially within those cantons with a high number of methadone treatments and a high proportion of known years of onset. Figure 5 shows the time trends of standardized incidence estimates in all cantons of Switzerland. The bold line indicates the best incidence estimates using year of onset. Incidence trends were similar between almost all cantons with a clear peak in the very early 90s, except for the canton of St. Gallen (SG). The 2 thinner lines depict the incidence trends, calculated by year of birth, shifted by 21 years. Even though the overall trend of all 3 lines is similar within cantons, the trends calculated by year of birth generally show a less pronounced peak.

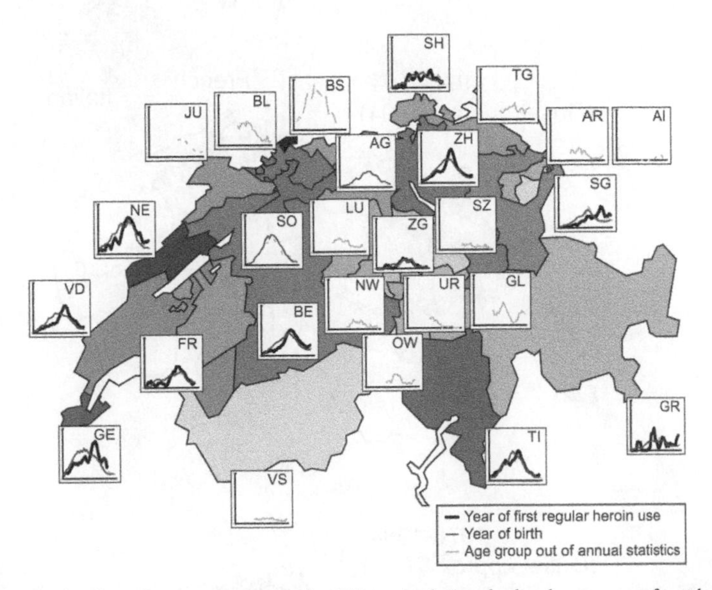

Figure 5. Standardized incidence estimates by the GIF method. Standardized estimates of incidence of regular heroin use per 1,000 population and its approximation by using year of birth or 5-year age group categories from annual tables of all 26 cantons of Switzerland, 1970–2005. Gaps in lines derive from missing annual tables in the National Statistic of Methadone Treatment between 1999 and 2006 [7].

Figure 6 displays approximated incidence trends over time in each canton of Switzerland. Darker colors represent higher incidence rates. Using estimates of the number of affected individuals born between 1962 and 1969 and initiating regular heroin use around 1986, we found 5 areas–each comprising 1 or more cantons–with higher incidence rates. These regions were spread over all linguistic areas of Switzerland. Over the years, differences in incidence rates between cantons disappeared and heroin incidence now seems to have stabilized at a comparatively low and equal level throughout Switzerland.

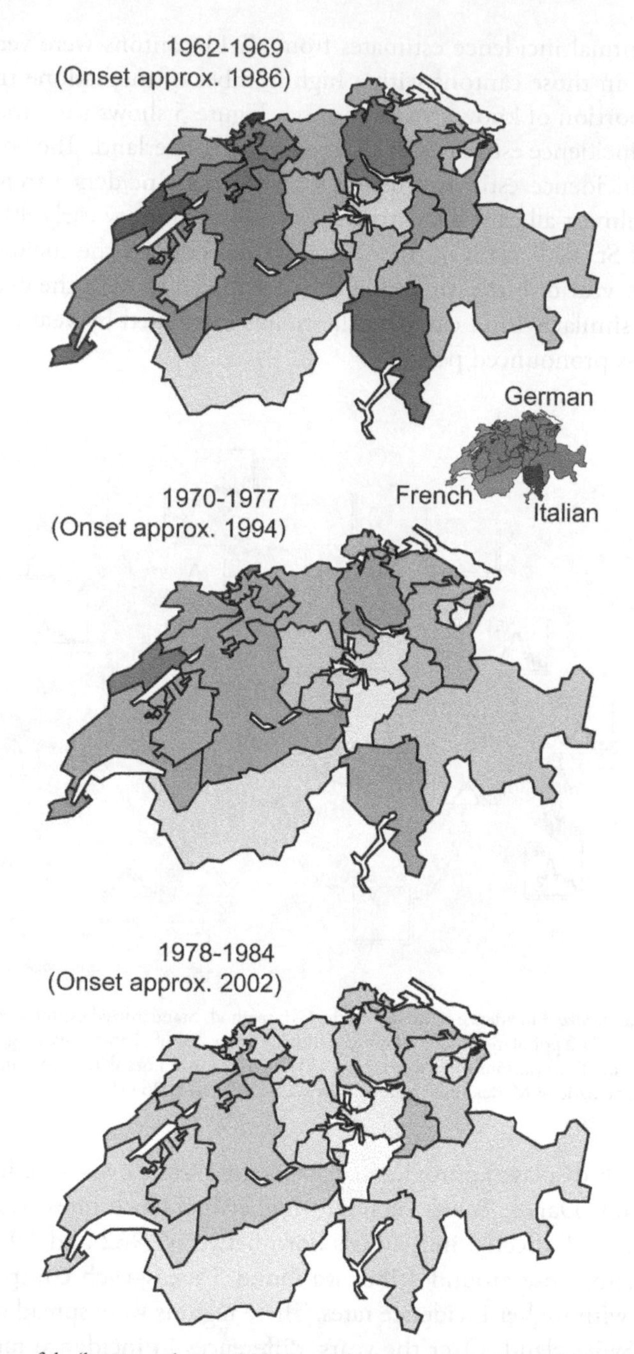

Figure 6. Evolution of the 'heroin epidemic.' Evolution of the 'heroin epidemic' in 3 phases using approximated incidence trends of regular heroin use per 1,000 population during 2 decades in all 26 cantons of Switzerland. The darker the area the higher the affected birth cohorts in the respective canton, using 5-year age group categories from annual statistic tables between 1999 and 2006.

Discussion

This study has shown that the GIF method allows us to estimate the incidence of regular heroin use even in small areas or with incomplete data. By joining long-term treatment register data from 11 cantons, we found that in Switzerland one half of regular heroin users had entered substitution treatment for the first time within two years of onset. Since the 11 cantons included in the study provide approximately 70% of all opioid substitution treatments in Switzerland, we estimate that in the whole of Switzerland, the incidence of regular heroin use peaked with about 3,675 new users in 1990, and that about 1,000 individuals began regularly using heroin per year from 2001 onwards.

In all cantons, the mean age of onset of regular heroin use was about 21. Therefore, supposing a constant age of onset of 21 years and applying the GIF method we estimate that for the whole of Switzerland the birth cohorts between 1965 and 1968 were most affected with approximately 2,800 (≈2,000/0.7) individuals initiating regular heroin use.

Using year of birth and assuming onset of regular heroin use at the age of 21 years led to reasonably good approximations of incidence estimates obtained by real age of onset. If we suppose that any short peak of heroin incidence will affect young people at around the age of 20, it follows that birth cohort incidence curves will be less steep and smoother than year of onset incidence curves. The results from different Swiss cantons showed some support for this assumption, with birth cohort trend estimates peaking a few years earlier at lower levels.

Given the relatively high proportion of missing data on age of onset, the variation in the proportion of missing data between cantons and the fact that the mean age of those patients with missing data on age of onset was about 3 years higher than that of patients with available data, we must be careful when comparing incidence trends between cantons. If data concerning age of onset is of poor quality, using year of birth and calculating incidence by birth cohort, may lead to better estimates. Among our incidence estimates, the canton of St. Gallen is a good example, where a rather pronounced difference between birth cohort estimates and year of onset estimates point to possible weaknesses in the quality of data on age of onset. However, even if birth cohort estimates are more robust, they do not necessarily correspond fully to onset incidence trends.

In almost all cantons with elevated incidence levels, the birth cohorts of the late 60s and the early 70s were most affected by heroin use. Due to the small size of some cantons and some data of debatable quality, we think that grouping estimates into 7-year periods is more appropriate than speculating about the 'irregularities' of estimated trends. Despite this low time resolution, our representation of the spread of the 'heroin epidemic' in its spatial and temporal dimensions is still meaningful. It clearly reveals that there were several regions with higher incidence rates during the 80s. These 'centers' are distributed throughout all linguistic regions of Switzerland. This might be surprising since open drug scenes have only been known to exist in the German part of Switzerland. Thus, we have no indication of whether the 'heroin wave' begun in the eastern or in the northern part of Switzerland, nor whether it spread from a centre–for example, from Zurich, with its more prominent open drug scenes. Over the last 2 decades, the former spatial differences seem to have disappeared and the incidence of regular heroin use has stabilized at a comparatively low level throughout Switzerland.

Apart from the afore-mentioned limitations, our approach should not be understood as providing the 'best possible estimate' of the incidence of regular heroin use for each canton of Switzerland. In this paper, we neglected the number of opioid users in heroin-assisted treatment programs. This form of treatment exists in about half of all cantons in Switzerland, most often in the German part [10]. But as the proportion of opioid users in heroin-assisted treatment amounts only to about 10% of those in methadone treatment, this would not lead to substantially different results. A probably greater influence on estimates in specific cantons is the implicit assumption that there was no long-term drift of heroin users towards more urban regions. Short-term drifts are less probable because of the activities of cities with open drug scenes that aimed to 'repatriate' users of illegal drugs to their former living community (see for example a project established in 1993 in the city of Zurich [11]). However, a long-term shift may have led to an underestimation of incidence trends in smaller cantons, where trends are difficult to estimate due to small numbers. Taken together, we recommend that incidence estimates for each calendar year should not be made for areas with less than 300 patients in substitution treatment. For greater time-frames even about 100 patients in substitution treatment will be sufficient.

Conclusion

Using a simple method to estimate the incidence of heroin use, we were able to show that it is also possible to estimate incidence trends in relatively small areas and with incomplete data.

As to the spread of the heroin wave in Switzerland, in all cantons incidence peaked at the beginning of the 90s. Those cantons with higher incidence rates during the 80s were not necessarily those cantons where open drug scenes were present. Over the last 2 decades the former spatial differences in Switzerland seem to have disappeared, and the incidence of regular heroin use has stabilized at a comparatively low level throughout Switzerland.

Competing Interests

The authors declare that they have no competing interests.

Authors' Contributions

CN was the main contributor to the design of the study and did the statistical analyses. All authors participated in the execution of the study and the writing of the manuscript. All authors read and approved the final manuscript.

Acknowledgements

This work was funded by a grant of the Swiss Federal Office of Public Health (07.006985/204.0001-432). The Swiss Federal Office of Public Health had no further role in study design; in the analysis and interpretation of data; in the writing of the report; or in the decision to submit the paper for publication. We would like to acknowledge the Swiss Federal Office of Health for providing most of the data sets and the cantonal authorities who gave us permission to use them. We would also like to acknowledge Sharon Arpa for improving the English.

References

1. De Angelis D, Hickman M, Yang S: Estimating long-term trends in the incidence and prevalence of opiate use/injecting drug use and number of former users: back-calculation methods and opiate overdose deaths. Am J Epidemiol 2004, 160:994–1004.
2. Hickman M, Seaman S, de Angelis D: Estimating the relative incidence of heroin use: application of a method for adjusting observed reports of first visits to specialized drug treatment agencies. Am J Epidemiol 2001, 153:632–641.
3. Ravà L, Calvani MG, Heisterkamp S, Wiessing L, Rossi C: Incidence indicators for policy-making: models, estimation and implications. [http://www.unodc.org/pdf/bulletin_2001-01-01_1.pdf]. Bulletin on Narcotics United Nations, New York; 2001, LIII(1 and 2):135–155.
4. Law MG, Lynskey M, Ross J, Hall W: Back-projection estimates of the number of dependent heroin users in Australia. Addiction 2001, 96:433–443.
5. Nordt C, Stohler R: Incidence of heroin use in Zurich, Switzerland: a treatment case register analysis. Lancet 2006, 367:1830–1834.

6. Nordt C, Stohler R: Estimating heroin epidemics with data of patients in methadone maintenance treatment, collected during a single treatment day. Addiction 2008, 103:591–597.
7. Swiss Federal Office of Public Health: Nationale Methadonstatistik [National Statistic of Methadone Treatment]. [http://www.nasuko.ch/nms/db/index.cfm].
8. Gervasoni JP, Dubois-Arber F, Benninghoff F, Spencer B, Devos T, Paccaud F: Evaluation of the Federal measures to reduce the problem related to drug use. [http://www.iumsp.ch/Unites/uepp/files/tox2_en.pdf]. Second synthesis report 1990–1996. Abridged version Institut universitaire de médicine sociale et préventive, Lausanne; 1996.
9. Petitjean S, Dürsteler-MacFarland KM, Strasser H, Wiesbeck GA: Das Phänomen einer alternden Patientenpopulation [The phenomenon of an aging patient population]. [http://www.gesundheitsdienste.bs.ch/kaed_06-02_nr8_methadon_evaluation.pdf]. Forschungsgruppe Bereich Abhängigkeitserkrankungen der Universitären Psychiatrischen Kliniken, Basel; 2006.
10. Swiss Federal Office of Public Health: Heroin assisted treatment (HAT). [http:/ / www.bag.admin.ch/ themen/ drogen/ 00042/ 00629/ 00798/ 01191/ index.html?lang=en].
11. City Police of Zurich: Vermittlungs-und Rückführungszentrum VRZ [Centre of Placement and Repatriation]. [http:/ / www.stadt-zuerich.ch/ content/ pd/ de/ index/ stadtpolizei_zuerich/ ueber_uns/ region_west/ kommissariat_mobil_west/ die_wichtigsten_aufgaben.html].

Substance Abuse, Treatment Needs and Access Among Female Sex Workers and Non-Sex Workers in Pretoria, South Africa

Wendee M. Wechsberg, Li-Tzy Wu, William A. Zule, Charles D. Parry, Felicia A. Browne, Winnie K. Luseno, Tracy Kline and Amanda Gentry

ABSTRACT

Background

This study examined cross-sectional data collected from substance-using female sex workers (FSW) and non-sex workers (non-SW) in Pretoria, South Africa, who entered a randomized controlled trial.

Methods

Women who reported alcohol use and recently engaging in sex work or unprotected sex were recruited for a randomized study. The study sample (N = 506) comprised 335 FSW and 171 female non-SW from Pretoria and surrounding areas. Self-reported data about alcohol and other drug use as well as treatment needs and access were collected from participants before they entered a brief intervention.

Results

As compared with female non-SW, FSW were found to have a greater likelihood of having a past year diagnosis of alcohol or other drug abuse or dependence, having a family member with a history of alcohol or other drug abuse, having been physically abused, having used alcohol before age 18, and having a history of marijuana use. In addition, the FSW were more likely to perceive that they had alcohol or other drug problems, and that they had a need for treatment and a desire to go for treatment. Less than 20% of participants in either group had any awareness of alcohol and drug treatment programs, with only 3% of the FSW and 2% of the non-SW reporting that they tried but were unable to enter treatment in the past year.

Conclusion

FSW need and want substance abuse treatment services but they often have difficulty accessing services. The study findings suggest that barriers within the South African treatment system need to be addressed to facilitate access for substance-using FSW. Ongoing research is needed to inform policy change that fosters widespread educational efforts and sustainable, accessible, woman-sensitive services to ultimately break the cycle for current and future generations of at-risk South African women.

Background

South Africa has one of the highest levels of alcohol consumption per adult drinker in the world [1]. In 2000, estimates indicated that alcohol use contributed to 7% of disability adjusted life years lost in South Africa, ranking third out of 17 risk factors studied [2]. Among patients in specialized substance abuse treatment centers, alcohol is the primary substance of abuse reported in eight of the nine South Africa provinces, with the exception of the Western Cape where methamphetamine is the primary substance of abuse reported at treatment admission [3].

The World Health Organization's Gender, Alcohol and Culture: An International Study project [4] has increased attention on the need to study gender differences in drinking and differential responses that might be useful in addressing problems related to alcohol use. In South Africa, research has shown that females drink less alcohol (by volume) and less frequently than their male counterparts [5]. Nonetheless, estimates of alcohol use among South African females indicate that approximately 30% are alcohol drinkers [1] and roughly a third of both male and female drinkers drink at risky levels over weekends [6]. One in 10 women surveyed for the Demographic Health Survey (1998) had experienced symptoms of alcohol problems (scoring 2 or higher on the CAGE assessment) during her lifetime. Women who are poor and who lack education were significantly more likely to report lifetime alcohol problems [6]. Data suggest an increase in lifetime drinking among young, Black African males and females; and that women may use alcohol and other drugs as a way to cope with current or past life stressors [7,8]. Furthermore, many poor young South Africa women conduct sex work in order to support their families, and they report that using alcohol and other drugs helps them to solicit clients and overcome their shyness [9]. Another indicator of alcohol abuse among South African females is the extremely high prevalence of fetal alcohol syndrome among South African children in several communities [10-12].

A systematic review of FSW studies around the world reported a risk of sexually transmitted infections (STIs) and HIV, but none mentioned the risk associated with alcohol or other drug abuse [13]. Very few studies have considered the substance abuse or treatment needs of FSW [14]. No studies to date have compared the severity of substance use disorders in FSW to non-SW to adequately understand their alcohol and other drug dependence, as well as possible treatment needs. These factors reinforce the critical need to reach vulnerable women to understand the differences between them and to inform intervention and treatment that focus on individualized and gender-specific issues.

A major international review of substance abuse interventions highlighted brief treatment specifically as one intervention that is likely to be effective in reducing the burden of alcohol abuse [15]. However, despite the need for treatment, females are underrepresented in substance abuse treatment facilities, with males comprising approximately 76% to 90% of treatment center patients in all nine South African provinces [3]. Black South Africans, both male and female, are also underrepresented in treatment facilities [16,17]. Efforts to reduce treatment barriers–such as street outreach, outreach in township areas, and transportation–have not been adequately adopted by the majority of treatment facilities, despite the fact that taking these steps could potentially make treatment services more accessible to disadvantaged populations, and especially to females [16].

Female substance abusers, however, are not a homogenous group. In particular, FSW may represent a subpopulation that is particularly disadvantaged in terms of access to substance abuse treatment and other services. This population of women is also of particular relevance from a public health perspective because they are considered a core HIV transmitter group [18]. Among a study of predominately FSW in South Africa, almost 60% who used alcohol and other drugs were found to be HIV positive [19]. Despite the fact that many FSW abuse alcohol, research on this population in South Africa has tended to focus on sex risk, drug use and/ or violence, rather than examining substance abuse more broadly relative to other vulnerable females [20-22].

Consequently, this study aimed to (1) examine the characteristics, age of onset, and prevalence of substance abuse disorders (within the past year), including lifetime disorders, among a group of females in Pretoria who self-identified as FSW and those who self-identified as non-SW but had unprotected sex and also use alcohol; (2) examine perceived substance use problems, the need for substance abuse treatment services, and access; and (3) determine differences between FSW and non-SW on lifetime use and current alcohol and other drug use and dependence.

Methods

Participants

Data for this study were obtained from a randomized controlled trial among females at high risk for HIV in Pretoria. Study participants were recruited over a 3-year period (June 2004 to June 2007) in Greater Pretoria, which includes the central business district, nearby residential areas, and surrounding townships. A variety of methods (e.g., street-based outreach, fliers, and peer advocates) were used to recruit participants from target communities and areas known for illicit drug activity and sex work, including daily rate hotels, informal settlements, weekly apartment dwellings, and shelters that were identified into sampling zones.

Eligibility criteria for the study included the following: being female, at least 18 years of age, reporting alcohol use on at least 13 of the past 90 days, reporting either trading sex for money or drugs in the previous 90 days or having unprotected sex in the previous 90 days, providing written consent to participate, and providing verifiable locator information for Gauteng Province.

Based on a quick field screener that asked the eligibility questions, females who met preliminary screening criteria in the field were referred to the project office where final eligibility was determined by repeating the screener and informed consent obtained for study participation. Appointments and transportation arrangements to the project office were made for all potential participants who met the preliminary eligibility criteria. Intake data collection began with a locator form to enable outreach staff to contact participants for subsequent assessments. Field staff then conducted urine drug screens for cocaine, cannabis, opiates, amphetamines/methamphetamine, and Ecstasy use. Breath alcohol testing was performed to determine the breath alcohol concentration at the time of the interview.

Following intake procedures, study participants were assessed by self-report at a two-part baseline interview occurring 2 to 4 days apart as well as at 3- and 6-month follow-up interviews. The consent process and additional forms and baseline interview were deemed to be too time-consuming at pretesting to be conducted in one session, and therefore they were split into two intakes. In addition, the second intake increased participation for the experimental stage of the study. Data collection was performed using paper-and-pencil interviews. Consent forms, instruments, and intervention cue cards were available in English and two local languages, Sotho and Zulu, that were translated and backtranslated by South African Medical Research Council staff. All study activities were approved by RTI International's Office of Research Protection and Ethics, and the Human Research Ethics Committee at the University of Witwatersrand in Johannesburg.

This study is based on cross-sectional analysis of 506 participants, with complete baseline data as of June 2007. Among these participants, 335 (66%) reported trading sex in the past 90 days and 171 (34%) reported having unprotected sex in the past 90 days.

Measures

The demographic variables included age (split between 18–25 years and 26–55 years), level of education (lower than 7th grade, 7th to 12th grade, higher than 12th grade), current marital status (single; involved but not living with partner; living with partner; married, separated, divorced, or widowed), and number of children (none, one, two or more). Participants were also asked if their residence or living space had running water (yes, no) and/or electricity (yes, no). Family history of alcohol or other drug abuse was also assessed (yes, no), as well as history of being physically abused (yes, no) and/or sexually abused (yes, no). Additionally, participants were asked whether they had ever been tested for HIV (yes, no); those who reported having ever been tested were asked if they had ever been informed they were infected with HIV (yes, no).

Sex worker status was assessed with the question, "Have you traded sex for drugs, money, food, clothing, shelter or any other goods in the past 90 days?" Participants who responded yes were coded as sex workers; participants who responded no were coded as non-sex workers.

Respondents were asked a series of questions about their lifetime use and age of first use of alcohol, tobacco, marijuana by itself, Ecstasy, crack, cocaine by itself, heroin by itself, marijuana and heroin mixed, cocaine and heroin mixed, Mandrax (a sedative similar to methaqualone) by itself, Mandrax and marijuana mixed, LSD, Rohypnol, and inhalants.

Alcohol and other drug use disorders (abuse and dependence) were assessed by two separate sections, used previously in the Women's II Health CoOp, asking participants specifically whether their symptoms/problems were related to alcohol or other drug use. Assessment items were consistent with the criteria specified by the DSM-IV [23], and overall showed acceptable to excellent psychometric properties, which are presented in Table 1. The following four substance abuse criteria, as defined by the DSM-IV, were assessed: (1) recurrent substance use resulting in failure to fulfill major role obligations, (2) recurrent substance use in physically hazardous situations, (3) recurrent substance-use-related legal problems, and (4) continued substance use despite persistent or recurrent social or interpersonal problems. Lifetime abuse was defined as ever meeting one or more of these four abuse criteria. Past-year abuse was defined as meeting one or more of these criteria in the previous year.

Table 1. Descriptive statistics and reliability information on the alcohol and drug abuse and dependence scales

Scale Item (When was the last time that...)	Mean	SD	Reliability
Overall Alcohol Abuse & Dependency	34.19	8.97	0.8
Overall Drug Abuse & Dependency	43.77	10.81	0.9
Alcohol Abuse	12.23	3.34	0.7
You kept using alcohol even though you knew it was keeping you from meeting your responsibilities?	2.70	1.28	
You used alcohol where it made the situation unsafe or dangerous for you, or where you might have been forced into sex or hurt?	2.98	1.23	
Your alcohol use caused you to have problems with the law?	3.43	1.03	
You kept using alcohol even after you knew it could get you into fights or other kinds of legal trouble?	3.12	1.21	
Alcohol Dependence	19.42	5.89	0.8
You needed more alcohol to get the same high or found that the same amount did not get you as high as it used to?	2.69	1.33	
You had withdrawal problems from alcohol like shaking hands, throwing up, having trouble sitting still or sleeping, or that you used any alcohol to stop from being sick or avoid withdrawal problems?	2.88	1.32	
You used alcohol in larger amounts, more often, or for a longer time than you meant to?	2.55	1.32	
You were unable to cut down or stop using alcohol?	2.43	1.32	
You spent a lot of time either getting alcohol, using alcohol, or feeling the effects of alcohol (high, sick)?	2.85	1.32	
Your use of alcohol caused you to give up, reduce, or have problems at important activities?	2.99	1.28	
You kept using alcohol even after you knew it was causing or adding to medical, psychological, or emotional problems?	3.04	1.30	
Drug Abuse	14.03	3.15	0.8
You kept using drugs even though you knew it was keeping you from meeting your responsibilities?	3.33	1.15	
You used alcohol or drugs where it made the situation unsafe or dangerous for you, or where you might have been forced into sex or hurt?	3.42	1.09	
Your drug use caused you to have problems with the law (police)?	3.71	0.79	
You kept using drugs even after you knew it could get you into fights or other kinds of legal trouble?	3.57	0.97	
Drug Dependence	23.39	6.50	0.9
You needed more drugs to get the same high or found that the same amount did not get you as high as it used to?	3.36	1.15	
You had withdrawal problems from drugs like shaking hands, throwing up, having trouble sitting still or sleeping, or that you used any drugs to stop from being sick or avoid withdrawal problems?	3.34	1.19	
You used drugs in larger amounts, more often, or for a longer time than you meant to?	3.26	1.21	
You were unable to cut down or stop using drugs?	3.15	1.27	
You spent a lot of time either getting drugs, using drugs, or feeling the effects of drugs (high, sick)?	3.36	1.16	
Your use of drugs caused you to give up, reduce, or have problems at important activities?	3.47	1.06	
You kept using drugs even after you knew it was causing or adding to medical, psychological, or emotional problems?	3.45	1.08	

The following seven substance dependence criteria, as defined by the DSM-IV, were assessed: (1) tolerance, (2) withdrawal, (3) using the substance in larger amounts or over a longer period than intended, (4) persistent desire or unsuccessful attempts to cut down or stop substance use, (5) spending a great deal of time obtaining or using the substance or recovering from its effects, (6) reducing or giving up important social, occupational or recreational activities because of substance use, and (7) continued substance use despite knowledge of a physical or psychological problems. Participants who met at least three dependence criteria in their lifetime were classified as lifetime dependence. Past-year dependence was restricted to individuals who met at least three dependence criteria in the year preceding the interview. For both abuse and dependence, the same logic was applied to alcohol and other drug use diagnoses.

The subset of participants who met the criteria for past-year alcohol or other drug abuse or dependence were also assessed on perceived alcohol and other drug problems, their knowledge of treatment programs, whether they had ever called a treatment program for information or counseling, whether they had received treatment in their lifetime and in the past year, their perceived need for treatment, and whether they wanted to go to treatment.

Statistical Analysis

Descriptive analyses were conducted on the complete sample and chi-square tests were conducted to determine the difference in demographics, socioeconomic status, history of abuse, family history of substance abuse, HIV testing and status, and substance use characteristics between the two groups of FSW and non-SW. Logistic regression models were used to identify the characteristics associated with past-year (recent or active) alcohol and other drug abuse disorders. Finally, the analysis examined whether the FSW were different from the non-SW in self-perceived substance use problems, use of substance abuse treatment, and perceived need for treatment.

Results

There was no significant difference in age between the FSW and non-SW; however, the FSW were slightly older (Table 2). Significant differences in education and current marital status were noted, although both groups were very similar in the number of children. Large differences were found in regard to living conditions, with less than 50% of the FSW reporting electricity and running water where they live. The FSW had significantly higher rates of reported physical and sexual abuse, although high prevalence rates pertain to both groups overall. Alcohol use onset was very similar between the two groups in regard to whether drinking began before or after age 18. Significant differences were found with regard to marijuana use, with two thirds of the FSW reporting use compared with only a third of the non-SW.

Table 2. Sociodemographic characteristics, by sex worker status

Characteristic, Column %	Overall N = 506	Female sex workers[a] N = 335	Non-Sex workers N = 171	χ^2 test[b] p-values
Age Group (years)				
18–25	47.4	44.8	52.6	0.094
26–55	52.6	55.2	47.4	
Years of Education				
Lower than 7th grade	15.4	21.5	3.5	<0.001
7th to 12th grade	76.9	74.3	81.9	
Higher than 12th grade	7.7	4.2	14.6	
Current Marital Status				
Single, without a main sex partner	22.9	34.3	0.6	<0.001
Not living with a main sex partner	48.3	37.0	70.2	
Living with a main sex partner	24.7	26.5	21.1	
Married, separated, divorced, or widowed	4.2	2.1	8.2	
Number of Children				
None	33.8	34.3	29.8	0.186
One	33.4	31.0	39.2	
Two or more	32.8	34.6	31.0	
Residence or Living Space with Running Water, Yes	77.7	69.3	94.2	<0.001
Residence or Living Space with Electricity, Yes	59.3	43.3	90.6	<0.001
Family History of Alcohol or Drug Abuse[c], Yes	70.2	66.0	78.4	0.004
History of Being Physically Abused[d], Yes	53.8	63.3	35.1	<0.001
History of Being Sexually Abused[e], Yes	30.8	37.6	17.5	<0.001
Age of First Alcohol Use				
Before 18	46.4	45.4	48.5	0.500
18 or older	53.6	54.6	51.5	
Age of First Marijuana, Dagga, or Ganja Use				
Never used	44.9	32.5	69.0	<0.001
Before 18	19.0	20.3	16.4	
18 or older	36.2	47.2	14.6	

[a] Female sex workers are defined as females who had traded sex for drugs, money, food, or other goods in the past 90 days.
[b] Pearson chi-square test
NS: $p > 0.05$
[c] Family history of alcohol or other drug abuse: Ever having problems with alcohol or other drug use by any biological family member of the respondent.
[d] History of being physically abused: Ever being physically hurt by striking or beating to the point that the respondent had bruises, cuts, or broken bones.
[e] History of being sexually abused: Ever being pressured or forced to participate in sexual acts against the respondent's will.

Table 3 presents lifetime substance abuse overall and by both groups of females. Significant differences were found with most drugs examined, with higher rates always among the FSW. The main substances that show prevalence, aside from alcohol, are tobacco (67%), marijuana (55%), and crack cocaine (13%).

Marijuana (Dagga) mixed with Thai white (i.e., heroin; 8%) and Thai white alone (7%) are the next most commonly used drugs, but use is significantly different between the two groups. There is little use of club drugs, such as Ecstasy (3.4%) and LSD (0.4%).

Table 3. Lifetime substance use, by sex worker status

Prevalence of Lifetime Substance Use, Column %	Overall N = 506	Female sex workers[a] N = 335	Non-Sex workers N = 171	χ^2 test[b] p-values
Alcohol[c]	100	100	100	.....
Tobacco	67.4	73.4	55.6	<0.001
Any drug[d]	58.1	71.9	31.0	<0.001
Marijuana/Dagga/Ganja	55.1	67.5	31.0	<0.001
Ecstasy	3.4	4.2	1.8	0.152[e]
Crack	12.8	18.8	1.2	<0.001
Cocaine	2.6	3.6	0.6	0.044
Heroin/Thai White	6.9	9.3	2.3	0.004
Dagga and Heroin/Thai White	8.1	10.4	3.5	0.007
Cocaine and Heroin/Thai White	0.6	0.9	0.0	0.554[e]
Mandrax	1.2	1.5	0.6	0.669[e]
Dagga and Mandrax	1.2	1.8	0.0	0.101[e]
LSD	0.4	0.6	0.0	0.552[e]
Rohypnol/Shabba	3.0	4.2	0.6	0.024
Inhalants	3.0	4.5	0.0	0.005
Injection drug use	0.6	0.9	0.0	0.554[e]

[a] Female sex workers are defined as females who had traded sex for drugs, money, food, or other goods in the past 90 days.
[b] Pearson chi-square test
NS: $p > 0.05$
[c] Alcohol use was defined as any use of a whole alcohol drink in the lifetime.
[d] Any drug use included the use of marijuana (Dagga or Ganja), Ecstasy, crack, cocaine, heroin (Thai White), Mandrax, LSD, Rohypnol (Shabba), or inhalants.
[e] Fisher's exact test was used due to expected cell counts < 5.

Table 4 presents the prevalence rates of substance abuse disorders overall and between the two groups of females. Significant differences were found, with the FSW having more lifetime and past-year alcohol and other drug disorders than the non-SW. Although a high proportion of the non-SW were classified as having lifetime or past-year alcohol use disorders, significant differences were found, with higher rates among the FSW in all the categories of alcohol use diagnosis. Compared with the non-SW, the FSW also had significantly higher prevalence of lifetime and past-year drug abuse and dependence.

In aggregate, the FSW had a higher prevalence of any past-year alcohol or drug use disorder (84% vs. 66%). Regardless of their sex-work status, most participants abused or were dependent on alcohol alone (41%) or abused or were dependent on alcohol and other drugs (36%) in the past year. The prevalence of past-year abuse or dependence on drugs only (without alcohol) was very low; less than 2% in each group.

A logistic regression model of past-year alcohol or other drug abuse disorder with both unadjusted and adjusted odds ratios is presented in Table 5. A diagnosis of past-year alcohol or other drug abuse disorder was more likely for an FSW if she had a family member with a history of alcohol or other drug abuse, if she had been physically abused, if she had used alcohol before age 18, or if she had a history of marijuana use.

Table 4. Substance use disorders among women, by sex worker status

Prevalence, Column %	Overall N = 506	Female sex workers[a] N = 335	Non-Sex workers N = 171	χ^2 test[b] *p*-values
Lifetime Alcohol Use Disorder				
Abuse[c]	73.3	80.6	59.1	<0.001
Dependence	62.3	69.3	48.5	<0.001
Abuse or Dependence	80.0	85.7	69.0	<0.001
Past-Year Alcohol Use Disorder				
Abuse[c]	68.4	75.2	55.0	<0.001
Dependence	57.5	64.2	44.4	<0.001
Abuse or Dependence	76.1	81.8	64.9	<0.001
Lifetime Drug Use Disorder				
Abuse[c]	37.5	48.4	16.4	<0.001
Dependence	29.4	39.4	9.9	<0.001
Abuse or Dependence	40.9	52.2	18.7	<0.001
Past-Year Drug Use Disorder				
Abuse[c]	32.6	41.5	15.2	<0.001
Dependence	27.1	35.8	9.9	<0.001
Abuse or Dependence	37.2	47.2	17.5	<0.001
Any Lifetime Alcohol or Drug Use Disorder	81.8	88.1	69.6	<0.001
Alcohol and Drug Use Disorders	39.1	49.9	18.1	<0.001
Alcohol Use Disorder Only	40.9	35.8	50.9	
Drug Use Disorder Only	1.8	2.4	0.6	
Any Past-Year Alcohol or Drug Use Disorder	77.7	83.6	66.1	<0.001
Alcohol and Drug Use Disorders	35.6	45.4	16.4	<0.001
Alcohol Use Disorder Only	40.5	36.4	48.5	0.009
Drug Use Disorder Only	1.6	1.8	1.2	0.723[d]

[a] Female sex workers are defined as females who had traded sex for drugs, money, food, or other goods in the past 90 days.
[b] Pearson chi-square test.
[c] Abuse was defined as meeting DSM-IV criteria for abuse regardless of the status of dependence.
[d] Fisher's exact test was used due to expected cell counts < 5.

Table 6 presents the data for females who met the criteria for past-year alcohol or other drug abuse disorders. Significant differences were found between the two groups with regard to perceived alcohol and other drug use problems, with the FSW being more likely to perceive that they have alcohol and other drug problems and being more likely to perceive that they have a need for treatment. Three fourths of the FSW reported a desire for treatment.

Table 5. Logistic regression of past-year alcohol or drug use disorder (N=506)

Correlates of Past-Year Alcohol or Drug Use Disorder	Unadjusted OR (95% CI)	Adjusted[a] OR (95% CI)
Sex Worker (df = 1)		
Yes	2.6 (1.7–4.0)***	1.8 (1.0–3.2)*
No	1.0	1.0
Age Group (df = 1)		
18–25	1.0 (0.6–1.5)	--
26–55	1.0	
Years of Education (df = 2)		
Lower than 7th grade	1.2 (0.4–3.1)	--
7th to 12th grade	0.8 (0.4–1.9)	
Higher than 12th grade	1.0	
Current Marital Status (df = 3)		
Single, without a main sex partner	2.2 (0.8–6.2)	--
Not living with a main sex partner	1.6 (0.6–4.0)	
Living with a main sex partner	1.9 (0.7–5.1)	
Married, separated, divorced, or widowed	1.0	
Number of Children (df = 2)		
One	0.8 (0.5–1.4)	--
Two or more	0.6 (0.4–1.1)	
None	1.0	
Residence or Living Space with Running Water (df = 1)		
Yes	0.8 (0.5–1.4)	--
No	1.0	
Residence or Living Space with Electricity (df = 1)		
Yes	0.5 (0.3–0.8)**	0.9 (0.5–1.5)
No	1.0	1.0
Family History of Alcohol or Drug Abuse (df = 1)		
Yes	2.3 (1.5–3.5)***	2.5 (1.6–4.1)***
No	1.0	1.0
History of Being Physically Abused (df = 1)		
Yes	2.6 (1.7–4.1)***	1.7 (1.1–2.8)***
No	1.0	1.0
History of Being Sexually Abused (df = 1)		
Yes	3.1 (1.8–5.4)***	1.8 (1.0–3.3)†
No	1.0	1.0
Age of First Alcohol Use (df = 1)		
Before 18	2.3 (1.5–3.6)***	1.9 (1.1–3.0)*
18 or older	1.0	1.0
Age of First Marijuana, Dagga, or Ganja Use (df = 1)		
Before 18	4.0 (2.0–8.0)***	2.5 (1.2–5.2)*
18 or older	3.5 (2.1–5.8)***	2.4 (1.3–4.2)**
Never use	1.0	1.0

[a] The adjusted model included variables listed in the third column.
OR = odds ratio; CI = confidence interval
† $p \leq 0.09$; *: $p \leq 0.05$; ** $p \leq 0.01$; *** $p \leq 0.001$.
P-values are based on the Wald chi-square statistic.

Knowledge regarding alcohol and other drug treatment programs was limited in both the FSW and non-SW groups. This lack of awareness may partially explain why only a very small number of females in both groups reported having tried but having been unable to enter treatment in the past year; only 2% had ever been in treatment. It should be noted that the small number of females (n = 10) who reported an unsuccessful attempt to enter treatment in the past year precludes meaningful analysis of the barriers they may have encountered.

Table 6. Perceived substance use problems and treatment use that met DSM-IV criteria for a past-year alcohol or other drug use disorder

Respondents' Perceived Problems and Use of Treatment Services, Column %	Overall N = 393	Female sex workers[a] N = 280	Non-Sex workers N = 113	χ^2 test[b] p-values
Perceived alcohol problems	41.0	45.7	29.2	0.003
Perceived drug problems	29.3	36.1	12.4	<0.001
Perceived alcohol or drug problem	55.4	63.5	35.7	<0.001
Knew of any alcohol or drug treatment program	18.6	19.3	16.8	0.529
Perceived a need for treatment	63.4	72.9	39.8	<0.001
Wanted to go to treatment	68.2	77.1	46.0	<0.001
Tried but unable to get into treatment in past year[c]	2.6	2.9	1.8	0.731[e]
Received alcohol or drug treatment in the lifetime[d]	1.8	2.5	0.0	0.200[e]

[a] Female sex workers are defined as females who had traded sex for drugs, money, food, or other goods in the past 90 days.
[b] Pearson chi-square test
[c] Included having ever called a drug treatment program for information or counseling, having ever received telephone counseling from a drug treatment program, or having ever consulted a traditional healer for drug treatment.
[d] Included the receipt of treatment from a detox, outpatient alcohol, outpatient drug, outpatient methadone, residential addiction, or jail/prison treatment program.
[e] Fisher's exact test was used due to expected cell counts < 5.

Discussion

This study adds to the growing knowledge base about alcohol and other drug use by highlighting key differences between FSW and non-SW in a specific region of the world. Previous studies have shown that females in this region become sex workers because they do not have other employment options and often support multiple family members [9,24]. In addition, typically boys are favored in families for completing education and girls often do not complete schooling.

However, substance abuse intersects with other risks, including sexual and physical violence. Substance use also may assuage a woman's sense of embarrassment in conducting sex work and become part of an everyday ritual, which may help to explain the greater use of alcohol among FSW and their later initiation of marijuana use. In the formative stage of this study, some women mentioned how alcohol use has helped them feel assertive in talking with men to solicit clients [9]. Similarly, research into drug use and HIV risk behavior among FSW in three South African cities (one being Pretoria) found that FSW used drugs to help them get into the mood for sex work and to engage in sex acts with strangers [21].

The analysis demonstrated that there are greater differences between FSW relative to non-SW in terms of their background and their substance use and dependency. The FSW appear to be poorer and living without many of the everyday comforts, such as electricity and running water, compared with over 90% of the non-SW who have these essential amenities in their homes. The unadjusted odds ratio of having no electricity held in the logistic model as a significant independent variable. The FSW reported a significantly greater history of both physical and sexual abuse. Most females started drinking at a similar age; although those who used other drugs also started using at similar ages. In addition, there were no significant differences in age. Education, however, was significantly different, with the FSW reporting lower levels of education, which is a key underclass issue for females worldwide, as more education often means greater employability.

Whether the lack of economic opportunities in South Africa for women leads them to sex work remains speculative, but it is clear that FSW use drugs at a greater rate. Moreover, gender inequality and employment opportunities for females continue to be problematic [25]. The greater use of alcohol and marijuana–and to some degree crack and other drugs–by the FSW relative to the non-SW may be related to the nature of sex work and their subsequent need to continue to use drugs because of dependency, which in turn may put them at greater risk for further victimization, impaired sex, and HIV.

A diagnosis of lifetime dependence and abuse also showed that the FSW experienced both problems related to alcohol and other drugs. Although almost half of the FSW perceived that they had an alcohol problem and a third believed that they had a drug problem, a greater proportion perceived a need for treatment and wanted to go to treatment; however, very few ever entered treatment because they did not know of any programs.

Research on barriers to substance abuse treatment services in South Africa has shown that Black African women experience multiple barriers as FSW or non-SW [16,26]. Studies conducted among treatment centers in Gauteng Province (which includes Johannesburg and Pretoria) between 2003 and 2004 found that only 36% of centers provided woman-focused and gender-sensitive treatment programs. In general, few facilities in Gauteng provide services aimed at addressing some of the barriers–such as funding for treatment, childcare, and programs that focus on the special needs of women–that prevent women from accessing, engaging, and being retained in treatment [27].

The fact that many of the females participating in this study were not aware of treatment services but were eager to receive treatment raises questions. Thus, a logical next step would be to help these women learn about treatment, tailor treatment programs to be sensitive to women's needs, and address their comorbid conditions (e.g., sexual and physical abuse) and contextual issues (e.g., childcare). By implication, this also raises the issue of who will care for their other children and extended family if these women are not earning an income as sex workers.

Limitations

One limitation of this study is that the sample is not a random selection but a targeted purposive sample of at-risk females in a specific geographic area in South Africa. Some females may use alcohol and other drugs to cope with violence and/or their HIV status or simply their lifestyle. The findings about alcohol and other drug use are not generalizable to all South African females or even to all FSW. However, the findings offer additional detail about substance use within the context of the similarities and differences between these two groups of females.

Another limitation is that although the study sought females who use alcohol, because the study criteria selected females who drank 13 out of the past 90 days, there is no comparison with females who do not drink. Nonetheless, interesting similarities and differences were found between FSW and non-SW, which raises important considerations when designing and implementing intervention and treatment strategies.

Conclusion

Health service providers in this region might consider how to better reach and treat females with alcohol and other drug problems [28]. Intervention efforts should also focus on outreach strategies to continue reaching childbearing FSWs and other vulnerable females. These efforts should also address the intersecting risks that females face in South Africa because of gender inequality, as many resort to sex work because of too few or no options [25,29].

The findings of this study show that FSW need and want services, but they may be a group that is unable to access services because of what they do to support their families and because services may not be readily available or welcoming because of the stigma associated with sex work. The data show that there is a need for treatment for this population and that barriers to access need to be addressed within the South African substance abuse treatment system.

More research is needed to determine the effects of the comorbid conditions that affect these females and, in turn, study outcomes. Areas for further research suggested by this study include a greater need to understand the factors that protect females who live in difficult circumstances from becoming sex workers, such as increased education and ways to assist in accessing treatment services. Moreover, additional research will help to inform policy change that fosters widespread educational efforts as well as sustainable, accessible services that are aimed at ultimately breaking the cycle for current and future generations of vulnerable South African women.

Competing Interests

The authors declare that they have no competing interests.

Authors' Contributions

WW conceived this study, directed the research, and revised the manuscript. LW conducted the analysis. WZ contributed to the analyses and writing. CP contributed to the writing. FB contributed to the background and writing. WL conducted the literature review and wrote the methods. TK completed the psychometrics for the revision and table revisions. AG helped with the literature revisions.

Acknowledgements

This work was supported by the National Institute on Alcohol Abuse and Alcoholism (NIAAA) under grant number RO1 AA14488. The interpretations and conclusions are solely those of the authors and do not necessarily represent the position of NIAAA or the U.S. Department of Health and Human Services. We wish to thank all of our field staff in Pretoria, South Africa, and the women participants who made this research possible. We also wish to thank Jeffrey Novey for editorial assistance.

References

1. Rehm J, Rehn N, Room R, Monteiro M, Gmel G, Jernigan D, Frick U: The global distribution of average volume of alcohol consumption and patterns of drinking. Eur Addict Res 2003, 9:147–156.
2. Schneider M, Norman R, Parry C, Bradshaw D, Plüddemann A: Estimating the burden of disease attributable to alcohol use in South Africa in 2000. S Afr Med J 2007, 97:664–672.
3. Plüddemann A, Parry C, Cerff P, Bhana A, Sanca PE, Potgieter H: Monitoring alcohol and drug abuse trends in South Africa (July 1996–December2007). SACENDU Research Brief 2008, 11(1):1–12.
4. Obot IS, Room R, (Eds): Alcohol, Gender and Drinking Problems: Perspectives from Low and Middle Income Counties. Geneva, Switzerland: Department of Mental Health & Substance Abuse; 2005.
5. Simbayi LC, Kalichman SC, Jooste S, Mathiti V, Cain D, Cherry C: Alcohol use and sexual risks for HIV infection among men and women receiving sexually transmitted infection clinic services in Cape Town, South Africa. J Stud Alcohol 2004, 65:434–442.
6. Parry CD, Plüddemann A, Steyn K, Bradshaw D, Norman R, Laubscher R: Alcohol use in South Africa: findings from the first Demographic and Health Survey (1998). J Stud Alcohol 2005, 66:91–97.

7. Sawyer KM, Wechsberg WM, Myers BJ: Cultural similarities and differences between a sample of Black/African and Coloured women in South Africa: convergence of risk related to substance use, sexual behavior, and violence. Women Health 2006, 43:73–92.
8. Morojele NK, Kachieng'a MA, Mokoko E, Nkoko MA, Parry CD, Nkowane AM, Moshia KM, Saxena S: Alcohol use and sexual behaviour among risky drinkers and bar and shebeen patrons in Gauteng province, South Africa. Soc Sci Med 2006, 62:217–227.
9. Wechsberg WM, Luseno WK, Lam WK: Violence against substance-abusing South African sex workers: intersection with culture and HIV risk. AIDS Care 2005, 17:S55–S64.
10. Viljoen DL, Gossage JP, Brooke L, Adnams CM, Jones KL, Robinson LK, Hoyme HE, Snell C, Khaole NCO, Kodituwakku P, Asante K, Findlay R, Quinton B, Marais A-S, Kalberg WO, May PA: Fetal alcohol syndrome epidemiology in a South African community: a second study of a very high prevalence area. J Stud Alcohol 2005, 66:593–604.
11. Centers for Disease Control and Prevention: Fetal alcohol syndrome–South Africa, 2001. MMWR 2003, 52:660–662.
12. May PA, Brooke L, Gossage JP, Croxford J, Adnams C, Jones KL, Robinson L, Viljoen D: Epidemiology of fetal alcohol syndrome in a South African community in the Western Cape Province. Am J Public Health 2000, 90:1905–1912.
13. Shahmanesh M, Patel V, Mabey D, Cowan F: Effectiveness of interventions for the prevention of HIV and other sexually transmitted infections in female sex workers in resource poor setting: a systematic review. Trop Med Int Health 2008, 13(5):659–679.
14. Hong Y, Li X: Behavioral studies of female sex workers in China: a literature review and recommendation for future research. AIDS and Behavior 2008, 12(4):623–636.
15. Babor T, Caetano R, Casswell S, Griffith E, Giesbrecht N, Graham K, Grube J, Grunewald P, Hill L, Holder H, Homel R, Osterberg E, Relm J, Room R, Rossow I: Alcohol: No Ordinary Commodity–Research and Public Policy. Oxford, England: Oxford University Press; 2003.
16. Myers B, Parry CD: Access to substance abuse treatment services for Black South Africans: findings from audits of specialist treatment facilities in Cape Town and Gauteng. South African Psychiatry Review 2005, 8:15–19.
17. Parry CD, Plüddemann A, Bhana A: South African Community Epidemiology Network on Drug Use (SACENDU) Update–Alcohol and Drug Abuse Trends:

July-December 2006 (Phase 21). [http://www.sahealthinfo.org.za/admodule/sacendumay2007.pdf]. 2007.

18. Rees H, Beksinska HE, Dickson-Tetteh K, Ballard RC, Htun Y: Commercial sex workers in Johannesburg: risk behaviour and HIV status. S Afr J Sci 2000, 96:283–284.
19. Luseno W, Wechsberg WM: Correlates of HIV testing in a sample of high-risk South African Women. AIDS Care 2009, 21:178–184.
20. Leggett T: Drug, sex work, and HIV in three South African cities. Society in Transition 2001, 32:101–109.
21. Parry C, Dewing S, Petersen P, Carney T, Needle R, Kroeger K, Treger L: Rapid assessment of HIV risk behavior in drug using sex workers in three cities in South Africa. AIDS Behav 2008, in press.
22. Wechsberg WM, Luseno WK, Lam WK, Parry CD, Morojele NK: Substance use, sexual risk, and violence: HIV prevention intervention with sex workers in Pretoria. AIDS Behav 2006, 10:131–137.
23. American Psychiatric Association: Diagnostic and Statistical Manual of Mental Disorders. 4th edition. Washington, DC: American Psychiatric Association; 1994.
24. Hunter M: The changing political economy of sex in South Africa: the significance of unemployment and inequalities to the scale of the AIDS pandemic. Soc Sci Med 2007, 64:689–700.
25. Wechsberg WM, Luseno W, Riehman K, Karg R, Browne F, Parry C: Substance use and sexual risk within the context of gender inequality in South Africa. Subst Use Misuse 2008, 43:1186–1201.
26. Myers B, Parry CD, Plüddemann A: Indicators of substance abuse treatment demand in Cape Town, South Africa (1997–2001). Curationis 2004, 27:27–31.
27. Myers B: Substance Abuse Treatment Facilities in Gauteng (2003–2004). Parow, South Africa: Medical Research Council; 2004.
28. Wechsberg WM, Luseno W, Ellerson RM: Reaching women substance abusers in diverse settings: stigma and access to treatment thirty years later. Subst Use Misuse 2008, 43:1277–1279.
29. Wechsberg W, Parry C, Jewkes R: Research and Policy Brief. Drugs, Sex, and Gender-Based Violence: the Intersection of the HIV/AIDS Epidemic with Vulnerable Women in South Africa–Forging a Multilevel Collaborative Response. [http://www.mrc.ac.za/policybriefs/sa_policybrief0808_w2.pdf], 2008.

Nowhere to Go: How Stigma Limits the Options of Female Drug Users after Release from Jail

Juliana van Olphen, Michele J. Eliason, Nicholas Freudenberg and Marilyn Barnes

ABSTRACT

Background

Drug and alcohol using women leaving prison or jail face many challenges to successful re-integration in the community and are severely hampered in their efforts by the stigma of drug or alcohol use compounded by the stigma of incarceration.

Methods

This qualitative study is based on individual semi-structured interviews and focus groups with 17 women who had recently left jail about the challenges they faced on reentry.

Results

Our analysis identified three major themes, which are related by the overarching influence of stigma: survival (jobs and housing), access to treatment services, and family and community reintegration.

Conclusion

Stigma based on drug use and incarceration works to increase the needs of women for health and social services and at the same time, restricts their access to these services. These specific forms of stigma may amplify gender and race-based stigma. Punitive drug and social policies related to employment, housing, education, welfare, and mental health and substance abuse treatment make it extremely difficult for women to succeed.

Background

Drug and alcohol use and abuse (hereafter referred to as drug use) have devastating effects on the lives of individuals and families, and the health of communities. In the past few decades, drug use has become heavily stigmatized, resulting in the enactment of increasingly punitive drug laws. Policies such as the federal ban on food stamps for those convicted of a drug felony and the "One Strike, You're Out" law that evicts tenants with criminal histories from public housing have disproportionately and adversely affected women, especially poor women, limiting their options for employment, housing, and education upon release. [1] Between 1980 and 2002, the U.S. jail population increased by 265% resulting in an unprecedented number of people being released from jail on a daily basis. [2] Men are still the majority of those incarcerated in correctional institutions, but since 1990 the number of incarcerated women has grown at nearly twice the rate of men. [3] Women make up an increasing proportion of jail inmates, reaching 12.7 percent of the population in 2005, compared to 10.2 percent in 1995. [4] Because of U.S. criminal justice, drug and other social policies, particularly those related to the "war on drugs," drug-related offenses account for the largest share of the increase in the number of female offenders. [5] This is especially true in California where between 1986 and 2002, the number of women in California prisons increased by 311%; [6,7] and between 1986 and 1995, drug offenses accounted for 55% of the increase in the number of women in California prisons. [8]

Rising incarceration rates have had a disproportionate impact on the health of women of color who are overrepresented in jails and prisons. [9] A Black woman is more than seven times as likely to spend time behind bars as a White woman. [10] In San Francisco County Jail (SFCJ), eight in ten women are women of color and more than half are African American, despite the fact that less than 8% of the city's population is African American. [11] Some studies have found that services within the criminal justice system may not meet the specific needs of women of color, compromising their effectiveness. [12] Incarcerated women have higher rates of health problems such as HIV, hepatitis C, [8,9,13,14] and other sexually transmitted diseases, recent and chronic substance use disorders, [15-18] and mental health problems [18,19] than the general population.

Whereas much of the public attention has focused on people entering and returning from prison, each year more than 9 million people return home from jails, facilities that house status violators, detainees awaiting adjudication, and those sentenced to less than a year. Jails are even less likely than prisons to have adequate health care, substance abuse treatment services, or vocational training. [20,21] Although the average length of stay in jail is only about 45 days, evidence suggests that even a brief incarceration disrupts the lives of individuals, families, and communities. [16] Despite the high rates of health problems among incarcerated women, few receive treatment in jail or the discharge planning or aftercare that could link them with needed services after release. [15,16] In addition, incarceration and other social policies often contribute to a downward cycle of substance use and mental health problems. For example, a brief stay in jail often results in the termination of one's benefits including Medicaid, leaving newly released inmates with reduced or delayed access to mental health services and substance use treatment. [22-24] The lapse in benefits can also mean gaps in prescription drugs, significantly affecting physical and mental health. Changes in welfare policy imposed new restrictions on women as a condition for receipt of benefits. For example, some states require women to abstain from drug use before getting benefits, [17] making it more difficult for women with substance use problems to get treatment especially given the dearth of programs to address the needs of women dependent on drugs. Incarceration can also lead to homelessness or a change in housing status, [15] thus increasing the risk of precariously housed women for victimization, exploitation, or forced return to an unsafe situation. [24] Conviction for drug felonies can negatively affect the ability to get student aid and improve one's status in life through education. [1] Finally, since incarceration is concentrated in poor communities, most people leaving jail return to communities with limited access to education, housing and jobs and high levels of poverty, racism, drugs, violence and health problems. [15,25]

Stigma and Incarceration

Both drug use and incarceration carry stigma for men and women, [26,27] but the degree of stigma is much greater for women because of gender-based stereotypes that hold women to different standards. [28] The stigma of drug use and incarceration may be additive, yet little research to date has explored the impacts of multiple burdens of stigma on formerly incarcerated women. Stigma refers to unfavorable attitudes, beliefs, and policies directed toward people perceived to belong to an undesirable group. Erving Goffman, widely credited for conceptualizing and creating a framework for the study of stigma, described stigma as "an attribute that is deeply discrediting within a particular social interaction" (p.3). [29] Link and Phelan developed a conceptualization of stigma that describes stigmatized individuals as those who are labeled and assigned negative attributes, set apart as not fully human, and treated negatively. [30] Stigma results in prejudice and discrimination against the stigmatized group, reinforcing existing social inequalities, particularly those rooted in gender, sexuality and race. According to Link and Phelan, those who are stigmatized can experience direct, structural or internalized discrimination. For example, a formerly incarcerated woman may be treated poorly by others, denied access to housing or employment because of her criminal history, or internalize feelings of worthlessness because of the lowered expectations of those around her. This stigmatization is likely to significantly influence the success of a woman's transition from jail to home, potentially limiting her help-seeking intentions and compromising her access to health care, drug treatment, employment and housing. For many of these women, their stigma stems from the intersecting categories of incarceration history, drug use, mental health status, gender, race/ethnicity or sexual orientation, making it difficult to attribute any particular stigmatization to a single category.

Stigma contributes to policies related to the treatment of drug users, for example leading to strict standards of abstinence or clients are discharged from treatment as "failures," where victim-blaming is common as well as moral judgments about the "weak wills" of people who are thought to "choose" drug use. Incarceration stigma is expressed through a punishment rather than rehabilitation approach to drug use, a view of drug users as "criminals," zero tolerance for any use (or relapse), and a disdain for therapeutic interventions or compassion for those with drug addictions. The resultant criminalization of drug use means that relapse to drug use is the primary reason for a revocation of parole and return to prison for women. [31] The stigma can be internalized by those who use drugs as shame and guilt [27] which may exacerbate mental health problems, increase the risk for relapse, and result in low self-esteem.

In this study, we interviewed nine individuals and conducted two focus groups with eight other women who had been released from jail within the last 12 months about the challenges faced by drug users leaving jail. This report describes women's perceptions of the difficulties they faced upon release and the factors that eased their transition from jail to home.

Methods

The purpose of this study was to better understand the experiences of women with drug problems returning home from San Francisco County Jail. Women were able to participate in the study if they had been released from jail within the last 12 months and if they had been incarcerated for offenses related to their drug use or if they reported having drug problems at the time of arrest. Nine interviews and two focus groups were conducted. Focus groups were used to supplement interviews, because many women perceive focus groups as less threatening because they reduce the power differential of researcher versus research participant. [32] IRB approval for this study was obtained from the San Francisco State University Office for the Protection of Human Subjects.

Participants were primarily recruited through public housing in the Western Addition, a predominantly African American and low-income neighborhood in San Francisco, and through Northern California Service League (NCSL), a community-based organization located near the jail that provides services including life skills training, referrals, and vocational training to formerly incarcerated individuals. Flyers about the study were posted in both locations. In addition, potential participants were identified through staff at NCSL and through one of the interviewers for this project who lived in the Western Addition. The two focus groups were held at the offices of the NCSL, and interviews were held at a time and place convenient to the participants. The focus groups and several interviews were conducted by the first author of this study and the rest of the interviews were conducted by the fourth author, who was at the time of this study an undergraduate student at San Francisco State University. Data were collected in the Fall of 2005 and Spring of 2006. The interview and focus group guides were semi-structured, with a prepared list of topics and questions related to pre- and post-release experiences, particularly related to drug use, access to housing and healthcare, employment, and family/relationship issues. However, interview and focus group participants were also encouraged to share their own stories and to engage in a meaningful conversation with the interviewer, or with other participants in the focus group and its facilitator.

Interviews were analyzed according to standard qualitative techniques. [33-35] This included assigning codes to meaningful segments of transcript text and recording memos to help make sense of the data and facilitate more abstract development of theories about the data. Based on prior research, a number of themes were identified before beginning analysis, and these themes guided the development of the questions. Themes included: pre- and post-release challenges, treatment by staff in jail and by service providers post-release, challenges finding drug treatment and health care, employment and housing discrimination, and social support. The text of the interviews and focus groups was examined for the presence of these and other emergent themes; themes that repeatedly emerged in interviews and those emphasized by the respondents as important are reported here.

Results

Participant Characteristics

The average age of the 17 women participating in the study was 40 years (range of 22–53). The majority of participants (n = 10) were African American; the remaining seven participants described themselves as African American/White (2); White (2); Native American (1); Filipina (1); and Asian (1). Most participants (13) reported they had been in jail at least twice in the last 12 months. Most had used alcohol, heroin, methamphetamine, crack, cocaine, and/or marijuana in the past 6 months, but also tried to quit using drugs in the past six months. Most had unstable living situations, including several who were homeless or living in a car or a shelter.

Women's Experiences of Stigma Related to Their Drug Use and Incarceration

The themes that many women repeated can be organized into three interrelated consequences of drug and incarceration stigma: basic survival, access to needed treatment services, and family and community reintegration. These findings suggest that stigma helps to keep the revolving door of relapse, recidivism, and incarceration spinning. In the words of two respondents:

> *If they don't want people going back to jail they need to build their spirits up instead of breaking their spirits down, and that's all they're doing. I mean, "You're always going to [go to jail]. You're always going to be nothing but a whore. You're always..." Ain't nothing wrong with being what I'm being if I'm not hurting anyone. I'm not a murderer. I'm not out here killing nobody. And if I sold drugs, I sure as hell wouldn't sell it to somebody if it was going to kill them. You know? I'm not crazy. I wouldn't want to live with that for the rest of my life. San Francisco needs to be more caring.*

> *They [the police] say, "She's a known crack head and a crack addict." A lot of them like take my clothes and arrested me for [being] under the influence, and I'm not even high. And take my property one place and me another. Take my*

money. Leave my–they even don't even take my clothes. They leave my shit on the streets.

Survival: Jobs and Housing

The first concern of most women coming out of incarceration or drug treatment is making a living. This section shares some of the respondents' experiences trying to find employment and safe housing. Several women described what they perceived as job discrimination. For example, one woman said:

> *When I tried to get a job, behind me being a felon, I was unable to get a lot of jobs that I applied for. Because they say be honest about your history. And you're honest. And then you end up lying anyways because you need the work. And they're not going to hire–I have been so institutionalized. Yeah, I got a great resume because it's all transferable skills. But who's going to really hire me? Who's really going to give me the opportunity to make like $50,000 a year unless I go to school for the next ten years and prove that I am like a new citizen or whatever? But in the meantime, between-time, how am I going to pay rent? You know what I mean? How I am going to take care of myself if I don't have the support?*

Those who reported that they had been able to get a job felt that the jobs didn't pay them enough to survive in San Francisco or that they did not provide the benefits they needed to address their health problems. Another discussed how the cost of living in San Francisco made it even more difficult for those like her who can not find a job that provides a living wage:

> *You can't get certain jobs because you're a felon. And that's what really sucks. You might have the qualifications to get this job. Just because you're a felon, I'm not going to give you this job where you can make money so you can survive in San Francisco. And that's the hard part. We have to make money to survive out here. But it's not cheap. We can't just be making nine, ten dollars an hour. It's not going to work.*

Sometimes, the challenges women faced in getting a legitimate job that pays a living wage forced them to choose between unpalatable options–for example, sex work or selling drugs:

> *I'm a convicted felon, I'm not eligible for other things. Like I'm a drug addict. I'm not eligible for Proposition 36 (Appendix) because I sold dope. Well, to me, prostituting was too demeaning and I was raped too many times, so I stopped doing it. Right? So I started selling drugs. I'm still a drug addict. It's not like I sold*

drugs to become a rich person or anything. I sold drugs to pay my rent. I paid it. I lived in a room that was $50 a day, which was $1,500 a month.

Many women reported that they were forced to return to unhealthy or unsafe housing environments after release from jail because they had no healthier alternatives:

[the shelter] is drug infested... a lot of dope and it's dirty. It's scabies, body lice, hepatitis, TB, crabs, you name it.

A few women who were able to stop using drugs after release emphasized that a safe environment post-release helped them to stop using drugs. Women searching for alternatives to homeless shelters reported that incarceration severely compromised their ability to get housing, either because of their own or their family member's drug convictions:

Tried to go to Project Homeless Connect and everything and stuff like that to get housing, and can't get housing because my husband's got a felony. You know, stuff like that keeps us from being able to be free and have a place. All because of a drug conviction that he had, you know. It's not fair. People that are doing murders, they're getting the places that we could have.

This particular woman said she had no intention to stay off drugs upon release because she had "nowhere to go"; she had been on a waiting list for public housing for more than three years.

Access to Treatment Services

The majority of women reported that they had difficulty finding or benefiting from services because of the stigma of drug abuse or incarceration. Although approximately half of the women reported having been abused as children and all reported having current drug problems, few of them had received counseling or support for these problems while in jail. Whereas some women said they were offered the opportunity to participate in a jail-based drug treatment program, those who did attend said the program did not help them to address deep-seated psychological problems, often related to growing up in an unhealthy environment. One woman reported being asked to leave a program because of her behavior:

(At the program), they ask, 'Well, why are you using drugs?'

'Well, it *makes me feel better.*'

'Well, why does it make you feel better?'

'Okay, this is what I'm going through.'

'Well, there's no excuse.'

'No, there's no excuse but it's my choice.'

They tell me I shouldn't talk back. Talk back? You asked me a question, I'm going to answer it... I don't expect them to kiss my ass. But what I do expect is them to be a little more compassionate and caring. People who do drugs it's because they have a misfunction in their lives... It's not because they want to hurt anybody. It's because they have a balance that's gone.

This woman reported that the staff member's questions made her feel stigmatized as a drug user. Several other women reported that some program staff (both inside and outside the jail) made them feel worthless because of their situations and their drug use:

That's the problem with programs. They think they're better than you because they quit using. Fine, you have a reward for that. But don't put down somebody that hasn't been able to stop yet. That's just the way I feel about it.

However, several women reported more positive experiences in post-release programs, such as the women's group operated through the Pre-trial Diversion Program. Women are often mandated to participate in these groups as a condition of their release from jail. One woman expressed what she gained from attending this group:

Yeah, you learn how to express yourself in different ways–culturally, mentally, physically......It helps a lot to see who is and who isn't being fairly treated. And then it helped me a lot to see that there was a lot of people like me that weren't treated very good as children. You know, that's the growing up in an unhealthy environment, so you learn there's more people that are like you.

This woman, who had experienced childhood abuse, did not receive any help in jail for these problems. Another woman spoke about having made contact with someone from an organization serving low-income and homeless HIV-positive individuals. This staff person made a difference in her life because of the unconditional support she offered:

And she told me–supporting me in doing the right thing wherever I was at, whether I was on drugs or I was clean and sober. She always encouraged me to do the right thing cause it's hard when you get out. You don't have no family. You don't have anyone who is really supporting you in doing the right thing because everybody expects you're going to do the wrong thing anyways, so you just go right back to what you know.

This quote also highlights the way that the low expectations of other people in their lives led to a self-fulfilling prophecy for many women. Indeed, most women described an almost immediate return to drug use after release. Only the few women who reported that they had been released directly to a drug treatment program were exceptions to this norm. For example, one woman described how her last release was different from the previous ones:

Previous times [I was released], cause I didn't have nowhere to go previous times. I said to myself, "Okay, now I'm going to go back outside and go back and do the same thing." Cause I have no family out here. I have nothing. I have nowhere to go, and I have no way legally, the right direction of where to go, especially the first day you get out. It's hopeless.

This woman was eligible for Proposition 36. Another participant, however, said she was ineligible for the Proposition 36 services because she had been charged with drug selling.

For many women, the lack of treatment alternatives upon release meant an almost immediate return to their former lifestyles:

You know, after you go back out there, you aren't even worried about the services. You don't even think about real services when you relapse... All's about here making some money. Get me a room for when I–if I get tired.... Because I'm on the block 24/7 getting some money so I can support my habit. You're not worried about getting benefits anybody has.

Well, for me the last time that I had got out it was difficult even though I had an exit plan. I still was homeless. The Exit Plan was–it sounded good, and it looked good on paper. But the reality of it was when I came home, I–I mean the place where I was living was a crack house, and I didn't want to go back to the environment. But being that I came out with no ID, nothing, no money, nothing, I went right back to selling dope. I went right back to my drug addiction... I ended up back in jail two weeks later behind my drug addictions.

This second quote illustrates that an exit plan (a plan detailing what someone will do upon release to ensure successful reintegration), without the stable supports of an income or a drug-free place to live, may carry little benefit to someone leaving jail. Even those women who acknowledged the importance of the internal motivation and desire to quit admitted that these good intentions were not sufficient in the absence of a safe environment:

[Getting off drugs] got to be something that they want. Because I know I've had good intentions 101 times getting out, and I swore that I was not going to get high. And I always ended up loaded because I didn't have a safe place to go to. I went right back to the same environment and I kept all the same friends. And it's like people, places and things. And it's all triggers. And it's just all bad. And if you know nothing but drug addicts, you're going to go and do drugs. It's just like second nature. You know, to me, it's like I know how to hustle and get money. I don't know how to get a job. I just barely learned how to fill out a application. So it's things like that.

For this woman, drugs were the only way of life that she knew and living without drugs would mean learning a whole new way of living. When asked what they did in the first 24 hours after release, most women reported immediately returning to familiar places and people and getting high:

When you get the fresh air (after release), you lose your mind anyway. It's like freedom, and then you go see this man and this woman or whoever it is, you know, you get caught up in the codependency of drug addiction, and you want to go and have sex or whatever you haven't done in a mighty long time. You want to go and do that. So I think it would be helpful...if you could come and get (me) when I get discharged.

Other women told similar stories about the pervasiveness of addiction. Participant comments about the challenges of gaining access to city services revealed both structural barriers such as lack of coordination among agencies and individual barriers related to their drug dependency. One woman with mental health problems talked about being unable to be housed in a shelter and being endlessly shuffled around the city, referred from one place to another because of her mental health problems. This woman said she was living under the freeways.

So until I can learn how to deal my emotions, they feel it's better for me not to be in a shelter. So I was diagnosed [with borderline personality disorder] and I'm waiting–I'm on SSI and pending right now. I'm waiting. Waiting. Waiting. I don't know what's going to happen, but I want to get the counseling. And I want to get the help I need. And just I'm still waiting. Referral here. Referral there. Instead of actions, they're just referring me and referring and referring me. And it's getting aggravating.

Family and Community Reintegration

Three respondents expressed a deep desire to reunite with their children, arguing that the need to be good role models for their children was what compelled them to quit using drugs.

When I came home, [my daughter] was almost two years old. And she came over to me and put her hand on me and looked at me and said, "Mommy, I love you." And the fight was on from there. What's helping me now to abstain–I don't care how I look because I'm really abstaining.

Those who expressed a desire to reunite with their children also said they lacked a safe place where they could live with their children. The predominant experience of respondents who were mothers was that their continued substance use was a significant barrier to reunification. Several women felt they needed to be given more choices, rather than be pressured to reunite with their children when they may not be ready to do so:

You have to sit down and you have to find out what her plan is and what she feels. And if the other people around her are telling her, 'You have to do this or you don't love your kid,' that's going to make it even worse...as long as she's in a forced situation... and you don't know which way to turn, and you don't know what to do, and if you fuck up this way, you're going to lose your kid permanently. And that's usually what happens.

Two women talked about giving up their children because they wanted the best for their offspring, as illustrated by the following quotes:

So I couldn't put him through that (feeling of abandonment). So when he asked me to stay (with his foster parents), I let him stay. I signed him over to them. I just couldn't–it's not that I don't love my kids. It's just that I felt they'd be better off where they were.

My son is fourteen, and I have two daughters, an eight-year-old and a five-year-old. In their whole lives, I've been incarcerated, in and out of their lives. I don't even know that it would be helping to even be a part of their lives right now because what if I do relapse? What if I can't get it together? I mean I'm so used to failing, it's like to believe that I can actually succeed, I might set myself up and fail first before I make it to the miracle.

These quotes reveal the difficult choices some women are forced to make in the face of the challenges they faced getting a job, housing, and recovering from drug use to achieve enough stability to become good mothers to their children. They also noted that there is little support to get clean and sober in their home communities.

Because everybody I know is stuck in the same situation, not nothing positive. It's always crack, nothing regards achieving or talking about a goal or what do they want in life. But it's always crack. When I feel down and out and depressed, and I be trying to talk about my problems, but they be acting like they hear but they talking about a pipe.

Many participants discussed the cumulative stress and social isolation caused by incarceration that ultimately disrupted family and community connections.

And when I was in jail, it was just like they were taking little pieces of me away. And it's like if I stayed there longer, I probably would have been just totaled. You know, just totaled. I don't know if I could have handled it. I mean, I've done six months. Nobody visiting me. Nobody giving me any money. Nothing. I was there, and it was hard. It was really hard.

When I went to [jail], a lot of shit happened. My brother got his throat cut. My niece, she took a pill and almost OD'd. When I got out of the pen, when I came home, I knew my grandmother was dying. It was a lot of things, my family, there was no unity.

Suggestions Generated by the Respondents

Some of the women offered suggestions for improving the plight of drug users coming out of jail. For example, some mentioned the need for meaningful discharge planning, and another woman, quoted below, suggested peer education.

I think that if we were to help one another in the recidivism of our own lives is to become like–be peer educators and learn how to do some type of intervention for women who are transitioning... so we start this process 30 days to your discharge date and learn, okay, you get the people who like really want to change and do something different.

Some women felt strongly that they needed sober housing, for example : "a safe environment, alcohol and drugs and toxic-free. When I say toxic, I mean [free] of the negativity." Yet other women stressed the importance of housing for women who are still using, illustrating the heterogeneity of the population:

I just don't feel that (being abstinent) should be part of the rules. A lot of people will want to get off, but still use while they're going. And will go and slowly be tired of it, because you're not going to get off drugs until you're tired of it. You're not–you can't get off because you're forced to. Because if you do, you're going to relapse. You're going to relapse hard.

One participant shared a positive experience with early release from jail to a program under Proposition 36:

Previous times (I was released) I didn't have nowhere to go. I said to myself, "Okay, now I'm going to go back outside and go back and do the same thing... Prop. 36 they put me into a program. It's transitional. It helped me out a lot. It gave me housing and stability. And I'm happy.

Some women reported participating in the San Francisco Pretrial Diversion Program where they were mandated to go to groups as a condition of their release from jail. Benefits of the groups included

...learning how to express yourself in different ways.

It helped me a lot to see that there was a lot of people like me that weren't treated very good as children.

Discussion

This study has several limitations. This was a convenience sample of women, who may differ from the general population of women released from San Francisco County Jail in significant ways. In addition, the small sample size did not allow comparisons in perspectives on reentry by race/ethnicity or other personal characteristics. Since this study was designed to better understand women's reentry experiences in San Francisco, it may not reflect the experiences of women in other jurisdictions who are returning to their communities. In addition, it is important to note that this study was designed to elicit women's pre-release and post-release experiences and the challenges they experienced trying to find city programs and services. This study did not attempt to ascertain the veracity of women's self-reports.

A central theme throughout women's narratives revolved around the double stigma of being a drug user and having a history of incarceration. Drug users or those involved in the illicit drug economy are considered "suspect populations," groups of individuals who are highly stigmatized. [36] Public response to such populations, including public policy, attempt to contain and manage these individuals by stigmatizing drug use as a deterrent strategy. [37] Indeed, laws enacted as part of the nationwide War on Drugs have resulted in harsh penalties for drug users, including the prohibition of people with drug-related felonies from getting government assistance such as public housing and federal financial aid to attend college. In most cases, violent felons still have access to these benefits. [38] Finding employment after release was also exacerbated by incarceration-related stigma.

Even those with transferable job skills said that employers were unwilling to hire them as soon as they learned of their criminal history. Some suggested they were better off lying to prospective employers about their criminal history, perhaps an important survival strategy given research that has shown that 65% of employers would not knowingly hire someone who is formerly incarcerated, regardless of the offense. [27,39]

Although all of the women who participated in this study had recently been released from jail, not prison, the stories they told demonstrated ways that even a short stay in jail can disrupt one's life. The participants faced multiple, interrelated problems after release from jail, including drug use and mental health problems, family problems, lack of safe housing or a job that pays a living wage, challenges finding needed services, and social isolation, all stemming at least in part from stigma and discrimination related to their drug use and history of incarceration. Most of the women left jail unprepared to meet these challenges, some of which are related to regulations and practices that have inadvertently made successful community reentry more challenging. Few of the women we spoke to had received pre-release planning that helped them to find stable housing, find services or a job, or reunite with family members. Even women (usually only those who participate in programs in jail) who received an exit plan found that the plan was of little practical use on the outside. Pre-release planning can be a challenge when a woman's length of stay in jail is unknown or may last only a few days, but if reentry planning begins at a person's point of entry into the criminal justice system it has potential to reduce substance use and improve community health and public safety. Prior to release, women should be linked with community service providers who can help address their needs (e.g., for housing, employment etc.) after release.

Policies also hindered women's successful reintegration after release. For example, housing policies [40,41] often made it more difficult for people returning from jail to find stable housing. Lack of options after release, particularly a safe place to live, for most women we spoke to propelled them back into criminal activity, most often including drug use and drug selling as a survival mechanism. In the face of limited options, including a lack of gender-specific programs that allow women to live with their children or that provide childcare services [42-44] many women may be forced to choose between continuing custody and care for their children and drug treatment. The complexity of women's lives after release suggests that interventions designed to facilitate successful community reintegration must provide a range of services including substance use and mental health treatment, outreach, behavioral skills groups, intensive case management, reproductive health services, and programs linking women to housing and job training. Ironically, the social conditions that impaired their re-integration after a jail stay were typically the same conditions that led to their drug use and incarceration in the first place.

Although the mental health status of all of the women participating was not known, the evidence presented here suggests that many were struggling with mental health problems in addition to substance use problems. Other investigators have found that the dually diagnosed are a distinct population, in need of specialized integrated substance abuse and mental health treatment. [45] Stigma may be both a consequence of mental health status and a contributing factor to mental health problems because internalized shame creates emotional distress.

In San Francisco, Oakland and in other cities nationwide, a successful campaign known as "Ban the Box" has resulted in the elimination of the box on a form that applicants are required to check if they have had a felony conviction in the past. [46] While employers may still check the criminal records of prospective employees, the elimination of the box levels the playing field so applicants with a criminal history are at least initially considered alongside the other applicants in the pool. More efforts such as these are needed to combat the systematic social exclusion that results from the stigmatization of drug users who become entangled with the criminal justice system.

Finally, women's stories clearly illustrated how they felt poorly treated because of their status as a drug user, and the challenges women face finding the services they need are compounded by the quality of their interactions with service providers. For many women, the perception that they were being treated as an inferior exacerbated emotional problems with which women already struggled and contributed to relapse to drug use. [47] Stigma also resulted in mistaken assumptions about a woman's needs or priorities during reentry. While reunification with children might be a priority for some women, some who shared their stories clearly did not want to be pressured to make an immediate choice that might not be right for them or their children. Future research should more systematically examine the ways in which stigma exacerbates the problems women face after release, and constrains the opportunities available to them. There is also a need for studies to investigate how stigma might be internalized, exacerbating mental health problems and contributing to drug relapse.

Finally, incarceration and drug use added to the burden of stigma already elicited by the gender and race/ethnicity of the participants in this study, categories with lower levels of power in our society. In addition, gender and race/ethnicity play a major role in shaping the opportunities that women drug users leaving jail face, demonstrating the importance of understanding the intersecting patterns of stigma that block women leaving jail from successful reentry.

Conclusion

In an effort to discourage drug use and reduce crime, elected officials instituted policies intended to punish and stigmatize drug users to serve as a deterrent to drug use. This report showed that for the women who are the actual targets of these policies, the real impact was often the reverse: punitive policies, lack of services and stigmatization encouraged a return to drug use, increased criminal activity and re-incarceration, and exacerbated individual and community health problems. Even when women were able to find services, the stigma of drug use and incarceration often affected the quality of the interactions, even with providers of therapeutic services. In the future, launching campaigns to reduce the intersecting stigmas of drug use, incarceration, gender, and race/ethnicity, drawing on strategies currently used to improve mental health outcomes [48,49] may enhance the effectiveness of reintegration services while also assisting women leaving jail to find the support they need for successful reintegration into their families and communities.

Competing Interests

The authors declare that they have no competing interests.

Authors' Contributions

JVO conceived of the project and carried it out, ME assisted with organization and interpretation of the data, NF provided guidance on study design, data analysis, interpretation and organization of the manuscript, and MB assisted with conceptualization and implementation of the project, conducted some of the interviews, and reviewed the paper.

Appendix

Proposition 36, the Substance Abuse and Crime Prevention Law act, was passed in California in 2000. It allows first- and second-time nonviolent offenders charged with drug possession the opportunity to receive substance abuse treatment instead of incarceration.

Acknowledgements

The project described was supported by Award Number P20MD000544 from the National Center On Minority Health And Health Disparities.

References

1. Allard P: Life sentences: Denying welfare benefits to women convicted of drug offenses. Washington, DC: The Sentencing Project; 2002.
2. Harrison PM, Karberg JC: Prison and Jail Inmates at Midyear: 2003 Bulletin. 381 Washington, District of Columbia: U.S. Dept of Justice, Office of Justice Programs, 382 Bureau of Justice Statistics 2004.
3. Richie BE: The social impact of mass incarceration on women. In Invisible Punishment: The collateral consequences of mass imprisonment. Edited by: Mauer M, Chesney-Lind M. New York: The New Press; 2002:136–49.
4. Beck AJ, Harrison PM: Prison and Jail Inmates at Midyear 2005 U.S. Department of Justice, Office of Justice Programs. (NCJ-213133). [http://www.ojp.usdoj.gov/bjs/abstract/pjim05.htm].
5. Greenfeld LA, Snell TL: Women Offenders. Bureau of Justice Statistics Special Report. Washington, D.C.: U.S. Department of Justice; 1999.
6. California Department of Corrections: Women in jail in 2004.
7. Mauer M, Potler C, Wolf R: Gender and Justice: Women, drugs and sentencing policy. Washington, D.C., The Sentencing Project; 2004.
8. Maruschak LM: HIV in prisons and jails, 2002. Bureau of Justice Statistics; 2004.
9. Tonry M: Malign Neglect–Race, Crime and Punishment in America. New York, NY: Oxford University Press, Inc; 1995.
10. Bonczar TB, Beck AJ: Lifetime Likelihood of Going to State or Federal Prison. Washington, DC: US Department of Justice:. Bureau of Justice Statistics Special Report NCJ 1600092; 1997.
11. U.S. Census Bureau: State and County Quick Facts. 2000. Data derived from Population Estimates, Census of Population and Housing; 2000.
12. Primm AB, Osher RC, Gomez MB: Race and ethnicity, mental health services and cultural competency in the criminal justice system: Are we ready to change? Comm Mental Health J 2005, 4:577–569.
13. Grella CE, Annon JJ, Anglin MD: Drug use and risk for HIV among women arrestees in California. AIDS and Behavior 1999, 4(3):289–294.
14. McClelland GM, Teplin LA, Abram KM, Jacobs N: HIV and AIDS risk behaviors among female jail detainees: Implications for public health policy. AJPH 2002, 92(5):818–825.
15. Richie BE: Challenges incarcerated women face as they return to their communities: Findings from life history interviews. Crime Delinquency 2001, 47:410–427.

16. Freudenberg N: Jails, prisons, and the health of urban populations: A review of the impact of the correctional system on community health. J Urban Health: Bulletin of the New York Academy of Medicine 2001, 78(2):214–235.
17. Freudenberg N: Adverse effects of US prison and jail policies on the health and well-being of women of color. AJPH 2002, 92(12):1895–1899.
18. Peters RH, Hills HA: Intervention strategies for offenders with co-occurring disorders: What works?. The GAINS Center for People with Co-occurring Disorders in the Justice System; 1997.
19. Teplin LA, Abram KM, McClelland GM: Prevalence of psychiatric disorders among incarcerated women. Arch Gen Psychiatry 1996, 53:505–512.
20. Jose-Kampfner C: Health care on the inside. In Health issues for women of color. Edited by: Adams D. Thousand Oaks, CA: Sage; 164–184.
21. Yasunaga A: The health of jailed women: A literature review. J Correct Health Care 2001, 8:21–36.
22. Nelson M, Deess P, Allen C: The first month out: Post incarceration experiences in New York City. New York, NY: Vera Institute of Justice; 1999.
23. Van Olphen J, Freudenberg N, Fortin P, Galea S: Community reentry: Perceptions of individuals returning from New York City Jails. J Urban Health 2006, 83(3):372–381.
24. Aidala A, Cross JE, Stall R, Harre D, Sumartojo E: Housing status and HIV risk behaviors: Implications for prevention and policy. AIDS and Behavior 2005, 9(3):251–265.
25. James SE, Johnson J, Raghaven C: "I couldn't go anywhere." Contextualizing violence and drug abuse: A social network study. Viol Against Women 2004, 10:991–1014.
26. Link BG, Struening EL, Rahav M, Phelan JC, Nuttbrock L: On stigma and its consequences: evidence from a longitudinal study of men with dual diagnoses of mental illness and substance abuse. J Health Soc Behav 1997, 38:177–190.
27. Luoma JB, Twohig MF, Waltz T, Hayes SC, Roget N, Padilla M, Fisher G: An investigation of stigma in individuals receiving treatment for substance abuse. Addict Beh 2007, 32:1331–1346.
28. O'Brien P: Making it in the free world: Women in transition from prison. Albany, NY: SUNY Press; 2001.
29. Goffman E: Stigma: Notes on the Management of Spoiled Identity. New Jersey: Prentice Hall; 1963.
30. Link BG, Phelan JC: Conceptualizing Stigma. Annu Rev Sociol 2001, 27:363–385.

31. Harm NJ, Phillips SD: You can't go home again: Women and criminal recidivism. J Offender Rehab 2001, 32(3):3–21.
32. Wilkinson S: "Focus groups a feminist method." Psychol Women Q 1999, 23:221–224.
33. Strauss A: Qualitative analysis for social scientists. Cambridge, Cambridge University Press; 1987.
34. Strauss A, Corbin RJ: Basics of qualitative research. 2nd edition. London, Sage; 1990.
35. Miles MB, Huberman AM: Qualitative data analysis. Thousand Oaks, Sage; 1994.
36. Beckett K, Sasson T: The politics of injustice: Crime and punishment in America. Thousand Oaks, CA: Pine Forge Press, Sage; 2000.
37. Hartwell S: Triple stigma: Persons with mental illness and substance abuse problems in the criminal justice system. Criminal Justice Policy Review 2004, 15(1):84–99.
38. Maurer M: Invisible Punishment Block Housing, Education, Voting. Focus Magazine 2003, 3–4.
39. Holzer H: What employers want: Job prospects for less-educated workers. NY: Russell Sage Foundation; 1996.
40. Freudenberg N, Daniels J, Crum M, Perkins T, Richie B: Coming home from jail: the social and health consequences of community reentry for women, male adolescents, and their families and communities. Am J Public Health 2005, 95(10):1725–1736.
41. U.S. Department of Housing and Urban Development: Screening and eviction for drug abuse and other criminal activity.
42. Copeland J: A qualitative study of barriers to formal treatment among women who self-managed change in addictive behaviors. Journal of Substance Abuse Treatment 1997, 14:183–190.
43. Finkelstein N: Treatment issues for alcohol- and drug-dependent pregnant and parenting women. Health Soc Work 1994, 19(1):7–15.
44. McMahon TJ, Winkel JD, Suchman NE, Luthar SS: Drug dependence, parenting responsibilities, and treatment history: Why doesn't mom go for help. Drug and Alcohol Dependence 2002, 65:105–114.
45. National Gains Center: Intervention strategies for offenders with co-occurring disorders: What works. Substance Abuse and Mental Health Services Administration (CSAT/CMHS) and the National Institute of Corrections; 1997.

46. Legal Services for Prisoners With Children: [http://www.prisonerswithchildren.org/news/banthebox.htm]. All of Us Or None Bans the Box in San Francisco.
47. Ahern J, Stuber J, Galea S: Stigma, discrimination, and the health of illicit drug users. Drug Alc Depend 2007, 88:188–196.
48. Corrigan P, Gelb B: Three programs that use mass approaches to challenge the stigma of mental illness. Psychiatr Serv 2006, 57(3):393–8.
49. Palpant RG, Steimnitz R, Bornemann TH, Hawkins K: The Carter Center Mental Health Program: addressing the public health crisis in the field of mental health through policy change and stigma reduction. Prev Chronic Dis 2006, 3(2):A62.

Evidence for Habitual and Goal-Directed Behavior Following Devaluation of Cocaine: A Multifaceted Interpretation of Relapse

David H. Root, Anthony T. Fabbricatore, David J. Barker, Sisi Ma, Anthony P. Pawlak and Mark O. West

ABSTRACT

Background

Cocaine addiction is characterized as a chronically relapsing disorder. It is believed that cues present during self-administration become learned and increase the probability that relapse will occur when they are confronted during abstinence. However, the way in which relapse-inducing cues are interpreted by the user has remained elusive. Recent theories of addiction posit that relapse-inducing cues cause relapse habitually or automatically, bypassing processing information related to the consequences of relapse. Alternatively, other theories hypothesize that relapse-inducing cues produce an expectation of the drug's consequences, designated as goal-directed relapse. Discrete discriminative stimuli signaling the availability of cocaine produce robust cue-induced responding after thirty days of abstinence. However, it is not known whether cue-induced responding is a goal-directed action or habit.

Methodology/Principal Findings

We tested whether cue-induced responding is a goal-directed action or habit by explicitly pairing or unpairing cocaine with LiCl-induced sickness (n = 7/group), thereby decreasing or not altering the value of cocaine, respectively. Following thirty days of abstinence, no difference in responding between groups was found when animals were reintroduced to the self-administration environment alone, indicating habitual behavior. However, upon discriminative stimulus presentations, cocaine-sickness paired animals exhibited decreased cue-induced responding relative to unpaired controls, indicating goal-directed behavior. In spite of the difference between groups revealed during abstinent testing, no differences were found between groups when animals were under the influence of cocaine.

Conclusions/Significance

Unexpectedly, both habitual and goal-directed responding occurred during abstinent testing. Furthermore, habitual or goal-directed responding may have been induced by cues that differed in their correlation with the cocaine infusion. Non-discriminative stimulus cues were weak correlates of the infusion, which failed to evoke a representation of the value of cocaine and led to habitual behavior. However, the discriminative stimulus–nearly perfectly correlated with the infusion–likely evoked a representation of the value of the infusion and led to goal-directed behavior. These data indicate that abstinent cue-induced responding is multifaceted, dynamically engendering habitual or goal-directed behavior. Moreover, since goal-directed behavior terminated habitual behavior during testing, therapeutic approaches aimed at reducing the perceived value of cocaine in addicted individuals may reduce the capacity of cues to induce relapse.

Introduction

One of the most insidious characteristics of cocaine addiction is its chronic relapsing nature [1]. It is believed that various types of cues (paraphernalia, drug-associated odors and sounds, availability of cocaine, drug-use partners, etc) present during cocaine self-administration become learned and increase the probability that relapse will occur when an abstinent user is confronted with these cues [1], [2]. However, the behavioral mechanism that underlies cue-induced relapse is poorly understood. Recent theories of the neural basis of addiction posit that habit formation is a necessary contributor [2]–[6]. Habits have been defined as automatic behaviors that are insensitive to manipulations of their consequences [7]. Given that cocaine addiction is associated with a high risk of relapse despite negative consequences of returning to drug use such as sickness, depression, or loss of employment [1], it is reasonable to hypothesize that cue-elicited relapse is a habitual behavior. An alternate hypothesis [8] suggests that cues elicit an expectation of drug which drives drug-seeking (a goal-directed action). Distinguishing between goal-directed and habitual responding can be accomplished by manipulation of the response outcome [7], cocaine. In animal research, either of two methods, satiation of the reward or pairing the reward with an unpleasant outcome such as sickness, reduce the reward's "value" to the animal [7]. If either of these methods reduces the number of responses emitted in order to earn the reward in the devalued relative to the normal valued reward group, then the behavior is interpreted as a goal-directed action. In contrast, if there is no difference between devalued and valued groups, the behavior is deemed a stimulus-bound habit.

While experimenter-administered psychostimulants such as cocaine have been shown to enhance the formation of habitual responding for food [9]–[11], few reports have investigated whether cocaine self-administration behavior is controlled by habit. Dickinson and colleagues have demonstrated that oral sweetened cocaine-seeking behavior can become habitual [12], as tested by pairing the oral solution with LiCl-induced sickness. However, this report utilized an oral sweetened-cocaine solution as the reward, not the long utilized intravenous cocaine self-administration paradigm [13]. Furthermore, human cocaine self-administration studies have suggested that the intravenous route of self-administration has markedly greater potential for abuse than oral self-administration [14].

A recent report by Norman and Tsibulsky [15] found that intravenous cocaine self-administration behavior was completely blocked by satiety, and thus a goal-directed action rather than habitual. They reported that responding on the manipulandum that produced drug delivery, i.e., cocaine-seeking behavior, while under the influence of the drug was binary. When the animal was under the influence of cocaine, but not satiated, it exhibited 'compulsive' responding. Yet, when cumulative cocaine infusions reached the animal's 'satiety' level, the animal did not respond. One interpretation of those findings, although not new, is that drug level determines the rate of responding [13], [16].

Thus, while drug binge behavior may be goal-directed, it is unknown whether cue-elicited relapse, which is under abstinent conditions, is a goal-directed action or a stimulus-bound habit. While the link between craving and relapse is not completely understood [17], cues that signal the availability of cocaine (and other drugs) are potent producers of craving for the drug in cue-reactive individuals [18]–[21]. We have previously shown that discrete discriminative stimuli (S^D) that signal the availability of cocaine stimulate responding in rats following thirty days of forced abstinence [22]. In the present experiment, we tested whether cue-elicited relapse is a goal-directed action or a stimulus-bound habit by manipulating the "value" of cocaine via explicit pairings of cocaine with LiCl-induced sickness. Control animals received LiCl treatments that were not paired with cocaine. If cue-induced responding is goal-directed, response rates of paired animals should be lower than unpaired animals. In contrast, if cue-induced responding is a habit, response rates between paired and unpaired animals should not differ. Given that cocaine self-administration behavior is goal-directed [15], we hypothesized that cue-induced responding following thirty days of abstinence is also goal-directed.

Results

The overall experimental schematic is presented in Figure 1(see Methods). Consistent with various paradigms [15], [16], [22], [23], all animals learned to self-administer cocaine, increasing daily response rates and drug consumed over weeks of extended training. Over three weeks of self-administration training and prior to LiCl treatments, rats in both groups increased self-administered cocaine ($F(2, 24) = 13.49$, $P<.0001$) to a daily consumption level of 28.978±0.493 mg/kg/day (LiCl paired: 28.66±0.44; LiCl unpaired: 29.297±0.907) and increased the number of responses/min/day ($F(2, 24) = 8.19$, $p<.05$) to an average of 7.049±2.814 (LiCl paired: 10.011±5.539; LiCl unpaired: 4.047±0.840) in the third week. Although no statistical difference was found between groups, one outlier animal in the paired group (average responses/min/day was 41.632) inflated the average daily responses/min in the paired group (average responses/min/day without this animal: 4.740±1.869). Furthermore, there was no difference in time spent self-administering cocaine in either group ($t(12) = 1.269$, $p>.05$), averaging 5.790±0.041 hours/day for both groups in the third week of training (paired: 5.739±0.045; unpaired: 5.840±0.066). Thus, animals increased consumption of cocaine as well as the number of responses emitted. The ratio of total responses over total earned rewards significantly increased over weeks of self-administration ($F(2,24) = 7.10$, $p<.01$) to an average level of 33.043±12.869 in the third week (paired: 46.850±25.310; paired without outlier rat: 22.824±9.415; unpaired:

19.235±3.676). Such increasing response:reward ratios are consistent with those hypothesized to be critical in habit formation [7], [24]. There was no group difference (p>.05) or week × group interaction (p>.05) for any measure. On the last day of training the unpaired group emitted 4.748±1.117 responses/min, self-administering 30.193±0.551 mg/kg of cocaine, and the paired group emitted 14.74±9.631 (5.202±1.573 without outlier animal) responses/min, self-administering 29.726±0.493 mg/kg of cocaine.

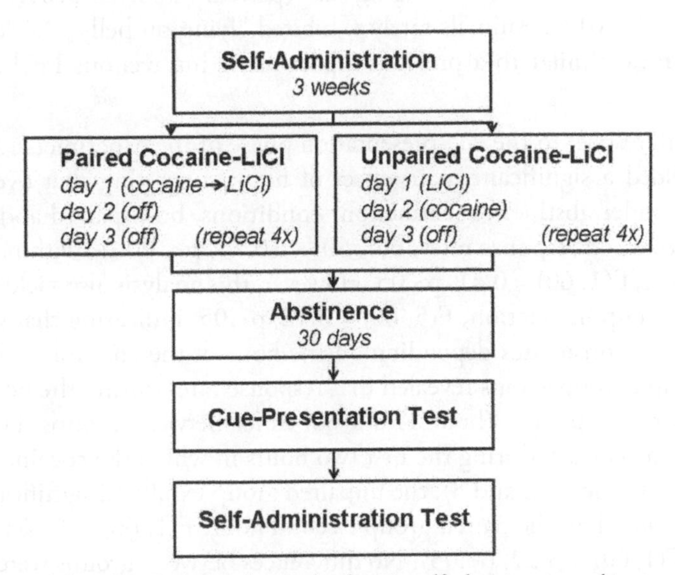

Figure 1. Experimental timeline. Following three weeks of cocaine self-administration, subjects were administered cocaine or LiCl on the same day (paired) or on separate days (unpaired). After thirty days of forced abstinence, animals were returned to the self-administration chambers for cue-presentation and self-administration tests.

Although all rats significantly decreased S_D reaction times (including both hits and misses) over weeks of training to 0.315±0.035 min/day in the third week of training (paired: 0.326±0.051; unpaired: 0.304±0.052), $F(2,24) = 22.51$, $p<.0001$, animals did not discriminate responding between S^D presence or absence, consistent with our previous study using the same schedule of reinforcement [22]. There was no significant group effect (p>.05) or week × group interaction (p>.05). On the last day of training, mean reaction time to the S^D for the paired group was 0.243±0.042 min and the unpaired group was 0.296±0.047 min.

Based on week three training data, we estimated the correlations of the S_D, lever, and operant chamber with the self-administered cocaine infusion. Rats responded on average 75.521 times out of an average 77.934 S^D presentations, giving the S_D the strongest correlation with the cocaine infusion at 0.969. In contrast, based on response:reward ratio's of 33.043 responses for a single infusion, the lever press had a 0.030 correlation with cocaine infusions. Furthermore, based on the maximum (80) number of 3.755-sec infusions of cocaine per day divided by 24 hours, the self-administration chamber had a 0.003 correlation with cocaine infusions. Thus, the S^D, but not the lever press or self-administration chamber, was highly correlated with the outcome, or infusion of cocaine.

During the devaluation phase of the experiment, there was no difference in total LiCl administered between groups ($t(12) = 0.096$, $p>0.05$), averaging 898.714±63.778 mg/kg/rat (paired: 905.143±109.751; unpaired: 892.286±74.613). Animals rarely exhibited "lying on belly", "sickness" behavior or diarrhea, similar to a previous report using intravenous LiCl administration [25].

With respect to the cue-presentation phase of the experiment, a mixed ANOVA yielded a significant main effect of hour, suggesting that over six hours of testing under abstinence (extinction) conditions, both paired and unpaired animals decreased response rates, $F(5, 60) = 10.75$, $p<.0001$, with no overall group difference, $F(1, 60) = 0.41$, $p>.05$. However, this analysis also yielded a significant hour × group interaction, $F(5, 60) = 3.12$, $p<.05$, indicating that groups differed in their response rates depending on the hour of the cue-test (Figure 2A). Post-hoc simple comparisons revealed that response rates during the hour in which S^D cues were not present (hour 1) did not differ between groups, $F(1, 60) = 0.60$, $p>.05$. In contrast, during the first two hours in which the cocaine-associated SD was present (hours 2 and 3), the unpaired group exhibited significantly higher response rates than the paired group (second hour: $F(1, 60) = 12.65$, $p<.001$; third hour: $F(1, 60) = 5.29$, $p<.05$). No differences between groups were found for the remainder of the hours.

These results suggest that cue-induced responding under abstinent conditions is decreased by prior devaluation of cocaine. However, it is possible that animals in the paired group extinguished to the return of the operant environment more quickly than animals in the unpaired group. To test this possibility, we analyzed the first 90 minutes of the cue-presentation session in fifteen minute bins. While the ANOVA yielded a significant main effect of bin, $F(5, 60) = 6.06$, $p<.001$ and no main effect of group, $F(1, 60) = 2.94$, $p>.05$, a significant bin × group interaction was revealed, $F(5, 60) = 3.30$, $p<.01$. Post-hoc simple comparisons revealed that response rates prior to S^D presentations did not differ between groups (Figure 2B). However, during the first fifteen minutes of SD presentations (bin 5), a significant difference between paired and unpaired animals emerged, $F(1, 60) = 4.70$, $p<.05$, and continued throughout the first thirty minutes of S^D presentations (bin 6), $F(1, 60) = 11.76$, $p<.01$. Thus, prior devaluation of cocaine produced significant differences in response rates between paired and unpaired groups selectively during S^D-exposure. Since groups contrasted in response rates only following the onset of S^D presentations, it is likely that the LiCl-paired devaluation of cocaine reduced S^D-induced responding rather than an acceleration of extinction responding upon returning to the operant environment. That is, the SD failed to excite responding in the paired group.

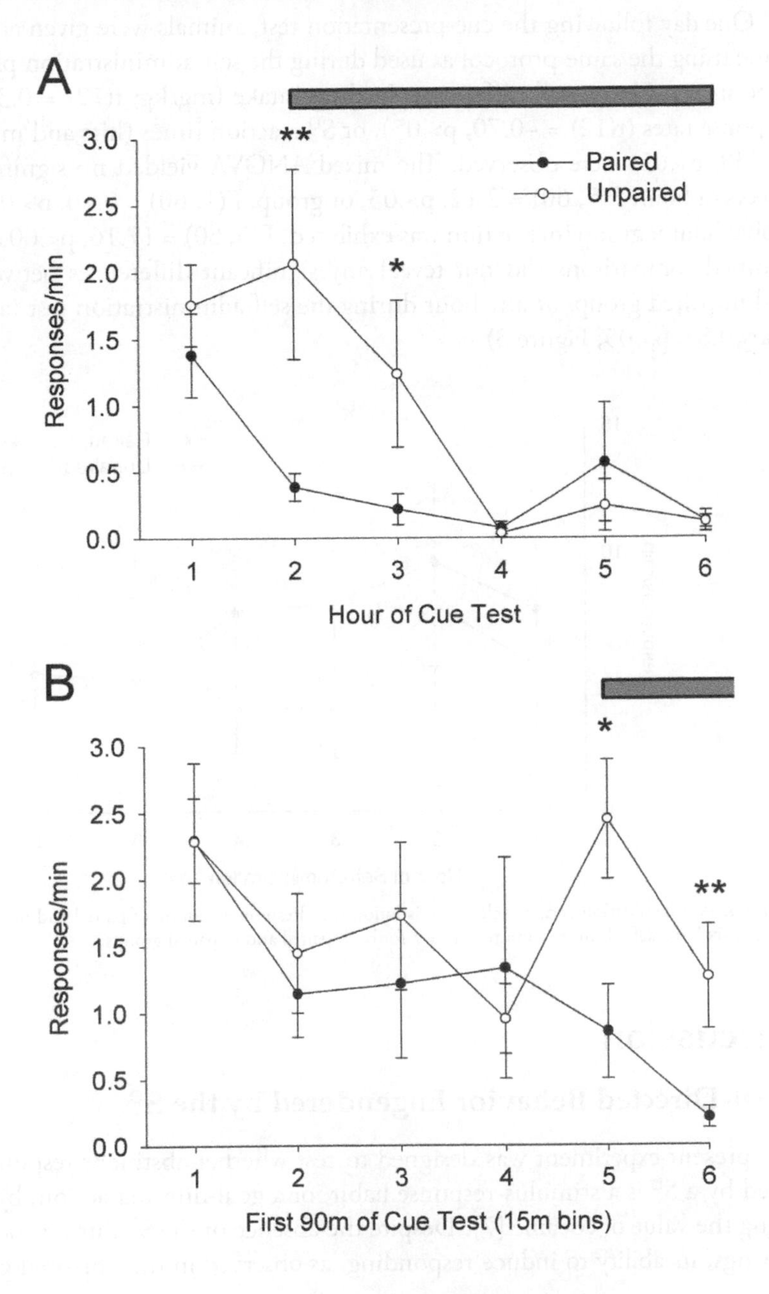

Figure 2. Cue-presentation test. A. Responses/minute of paired and unpaired animals over the six hour cue-presentation test. B. Responses/minute of paired and unpaired animals over the first ninety minutes of the cue-presentation test. Horizontal bars indicate period of variable 3–6 min interval S^D tone presentations, variably presented every 3–6 min. N = 7 for both paired and unpaired groups.

One day following the cue-presentation test, animals were given access to cocaine using the same protocol as used during the self-administration phase of the experiment. No overall differences in drug intake (mg/kg; $t(12) = 0.32$, $p>.05$), response rates ($t(12) = -0.70$, $p>.05$), or S^D reaction times (hits and misses; $t(12) = 0.39$, $p>.05$) were observed. The mixed ANOVA yielded no significant main effects of hour, $F(5, 60) = 2.12$, $p>.05$, or group, $F(1, 60) = 0.10$, $p>.05$. While a global hour × group interaction was exhibited, $F(5, 60) = 17.16$, $p<.001$, post-hoc planned comparisons did not reveal any significant differences between paired and unpaired groups at any hour during the self-administration test (all $F(1, 60)$ tests<0.87, $p>.05$; Figure 3).

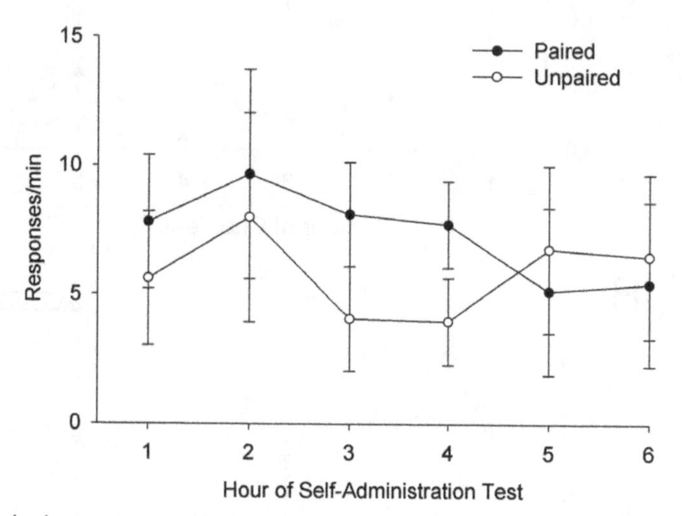

Figure 3. Post-devaluation cocaine self-administration test. Responses/minute of paired and unpaired animals over the six hour self-administration test. N = 7 for both paired and unpaired groups.

Discussion

Goal-Directed Behavior Engendered by the S^D

The present experiment was designed to test whether abstinent responding triggered by a S^D is a stimulus-response habit, or a goal-directed action, by manipulating the value of cocaine [7]. Despite the absence of the S^D during cocaine-LiCl pairings, its ability to induce responding, as observed in the unpaired group, was absent in the explicitly paired group. Reduced S^D-induced responding was likely due to explicit pairings of cocaine with LiCl because unpaired animals that received similar LiCl and cocaine exposure on separate days exhibited increased response rates upon S^D exposure. The observed SD-induced responding of unpaired animals was similar to animals with no LiCl exposure [22], demonstrating that LiCl did not affect the capacity of the S^D to engender responding. These data suggest that the value of the response outcome (cocaine) is not only evoked by the SD, but subject to manipulation following self-administration through cocaine devaluation.

In spite of three weeks of cocaine self-administration training, for paired animals, the SD predicted the aversive properties of LiCl that followed cocaine exposure during devaluation. In contrast, since LiCl was administered 24 hours following cocaine exposure in unpaired rats, the SD predicted an unaltered value of cocaine. Unpaired animals eventually reduced response rates to the level of paired animals during testing in extinction, indicating that the expectation of drug was not met when responses during S^D presentations did not produce cocaine infusions. The ability of the S^D to engender goal-directed responding may have been due to its unique, near 1.0 correlation with response contingent cocaine infusions and consequently, its unmatched prediction of reward in the current paradigm. Due to the close temporal proximity of the S^D with the infusion, cues that signal consumption of the outcome may remain sensitive to devaluation after overtraining, and by definition goal-directed as previously hypothesized [26].

Habitual behavior engendered by non-S^D cues

Equally noteworthy was the lack of differences in response rates between paired and unpaired groups during the hour prior to S^D presentations in the cue-presentation test. This finding suggests that responding in the absence of the SD and under abstinent conditions is habitual. Stimuli present during this time period were the self-administration chamber and lever. Unlike the S^D, these stimuli were poor correlates of cocaine infusions. That is, high response:reward ratios were observed during self-administration training, diluting the correlation of lever presses with cocaine infusions. Furthermore, during the self-administration phase of the experiment, animals lived in the self-administration chamber, spending 18 hours of each training day not self-administering. The weak correlation of these environmental cues with the cocaine infusion may be a factor in the development of habitual behavior [7], [27], [28] and the reason why the lever and self-administration chamber did not apparently evoke a representation of the devalued cocaine in the paired group and thereby influence response rates in the test hour prior to S^D presentations. Interestingly, the non-S^D cues were likely correlates of the tonically elevated levels of cocaine during training, similar to conditioned place preference. Since cocaine conditioned place preference is not blocked by LiCl [29], cues associated with the tonic levels of cocaine as well as those not strongly correlated with the outcome, may be particularly resistant to devaluation of the outcome (cocaine infusion in this case) and by definition, are effective producers of habitual behavior.

Studies have demonstrated that responding under the influence of cocaine is manifestly controlled by drug level [13], [15]. The present findings corroborate our earlier report in which animals' responding during self-administration did not discriminate the presence or absence of the S^D [22]. Given that our schedule of reinforcement precluded rats from attaining drug satiety, coupled with the animals' experience with long daily access and extensive training conditions, rats likely engaged in "compulsive" responding driven by and for cocaine [15], [23]. Despite the lack of evidence that responding was under stimulus control during self-administration, animals clearly learned a S^D-cocaine association in training, expressed as the paired group's reduced responding upon S^D exposure during the abstinent cue-presentation test. Given that cocaine taking is eliminated by satiety and thus a goal-directed action [15], it is not surprising, although not previously demonstrated, that cue-induced responding following thirty days of abstinence is also goal-directed. However, the finding that responding during the absence of the S^D was habitual suggests that over cocaine self-administration training, habitual behavior was latently developing, but was not expressed until induced by environment-cocaine associations under abstinence conditions.

Given the evidence that responding prior to and following the first hour of testing was goal-directed, one might question whether responding in the first hour was habitual. If so, the habit must have formed during self-administration training, but as noted above, there is no evidence to support this. Therefore, it must be considered whether behavior mechanisms other than habitual responding contributed to the observed response rates during the first hour of testing. For example, it has been reported that rats respond at higher levels during extinction from cocaine self-administration when the testing chamber is not the home cage [30]. Our experimental design involved the removal of rats from the self-administration/home chamber during abstinence and response rates were equally high in both groups when returned. This may be especially important because our prior investigation on S^D-induced responding yielded markedly diminished responding during the first hour of testing when animals were housed in the self-administration chamber over 30 days of forced abstinence prior to the cue-presentation test [22]. Regardless, the lack of difference between groups argues that responding in the first hour was not goal-directed. It is also possible that different methods of devaluation, such as single infusions of cocaine followed by single infusions of LiCl, could produce differential responding between groups during the first hour of testing. Finally, during training, different Pavlovian associations were likely formed between cocaine and the S^D from those formed between cocaine and the operant environment. The S^D may induce a more specific representation of the temporally proximal infusion, leaving the S^D more susceptible to evoking recollection of the devalued cocaine. In contrast, the operant environment and the lever, which were poor predictors of cocaine infusions, may have formed associations with tonically elevated levels of cocaine during self-administration training. If so, cues associated with tonically elevated levels of cocaine rather than the consumption of cocaine (the infusion) apparently do not evoke a representation of the value of cocaine. If these environmental cues had evoked a representation of cocaine's value, decreased responding would have been observed in the paired group during the first hour of abstinent testing relative to the unpaired group. Furthermore, this suggests that during devaluation training, paired animals learned that the infusion, rather than tonically elevated levels of cocaine, produces predicted sickness.

Differential Responding to Cues

Since responding during the absence of the S^D was habitual while responding during the presence of the S^D was goal-directed, abstinent responding may be driven by either of two behavioral mechanisms. Relapse is considered a process rather than a single event [8]. Throughout the process of relapse, an abstinent addicted cocaine user is likely to encounter a multitude of cues ranging from ambiguously to perfectly correlated with cocaine consumption. Depending on the strength of each cue's correlation with cocaine infusions, a spectrum of goal-directed behaviors and habitual behaviors may be engaged. However, since our results demonstrate that goal-directed behavior is able to terminate habitual behavior (i.e., hour 2 compared to hour 1), the ultimate determinant of relapse may be a goal-directed action. Moreover, habitual, or "absent minded" relapse [2] requires cocaine to be immediately accessible, whereas goal-directed relapse employs volition, necessary to manage the logistics (Where and how to get drug and paraphernalia?) and challenges (Take health, monetary, and punitive risks?) inherent in the process. Nevertheless, our results suggest that both mechanisms participate in the process of relapse.

Goal-Directed Cocaine Seeking

The notion that drug-seeking behavior is a goal-directed action is bolstered by behavioral economic analyses of animal self-administration behavior in which changes in drug choice are explicitly tied to changes in response cost [31], frequency of reward [32], dose per infusion [33], delay to reinforcement [34], infusion duration [35], feeding schedule [31], [36], probability of reinforcement [37], availability of alternative reinforcers [38], [39], and satiety [15].

In cocaine addicted individuals, choice paradigms pitting drug versus alternative rewards have revealed goal-directed actions to obtain drugs. The choice to self-administer cocaine over receipt of monetary reward depends on the dose the cocaine user will receive if he or she participates in the study [40]–[45]. In other words, addicted individuals do not work or "pay" for cocaine when the perceived expected value of the drug is reduced, implying a goal-directed process. Indirectly, investigations of self-administration behavior in cocaine addicted individuals have also revealed goal-directed self-reports. Specifically, cocaine addicted individuals self-report high ratings of "I want cocaine" while intravenously binging [46], [47] or in response to cocaine-associated cues [48], [49]. In some reports, cravings specifically for cocaine have been rated higher than nonspecific self-reports such as "rush", "high", or "excited" [50]–[53], but not in all cases [54]. It is interesting that when a cocaine-addicted individual is currently under the influence of cocaine by drug priming or self-administration, the choice to self-administer cocaine over monetary reward is nearly always cocaine [41], [42]. The lack of differences between paired and unpaired groups during the cocaine self-administration test may reflect that cocaine itself is a stimulus that can engender responding. Upon earning their first infusion of cocaine, animals may have entered into a "compulsive" state of responding [15], given that 1) the "priming threshold" that initiates responding for cocaine is less than one infusion of earned cocaine at the present dose, 2) responding does not cease until drug level reaches the "satiety threshold", and 3) our schedule of reinforcement precluded rats from attaining drug satiety. Furthermore, given that 1) discriminative responding to the SD tone is masked during self-administration but can be revealed during abstinent testing [22] and 2) the expression of habitual behavior is also masked during self-administration [15] but was revealed during abstinent testing in the present experiment, cocaine's presence during the self-administration test may have additionally masked the previously learned association between the cocaine infusion and sickness. Thus, the cocaine self-administration test, rather than indicating that LiCl sickness failed to devalue cocaine in the paired group, may produce results that are not akin to similar 'reacquisition' studies with natural rewards [7], [26].

Alternative Explanations

During the devaluation phase of the experiment, prior to infusions of LiCl, the paired animals received infusions of cocaine whereas the unpaired group did not receive identical infusions of saline. Thus, it is possible that nonspecific sensory properties of the infusion (i.e. increased venous pressure following pump activation) may have been associated with LiCl-induced sickness in the paired, but not unpaired group. However, this is unlikely to produce decreased response rates during the cue-presentation test for several reasons. First, except during self-administration and devaluation training, animals received saline infusions every fifteen minutes for over two months. Prior to devaluation training, over 2400 infusions of saline were administered per rat. The large number of exposures to these infusions, during which there were no consequences to the nonspecific sensory aspects of the infusions, would likely undermine any possible infusion-LiCl-induced sickness associations during devaluation, which were far fewer in number. Second, both the paired and unpaired animals received intravenous LiCl administration during devaluation training. If LiCl-induced sickness was to become associated with aspects of the LiCl infusion, it is likely to occur immediately preceding pump pressure. Yet since both groups received the same intravenous route of LiCl administration, both groups would have equally associated the nonspecific effects of the infusion pump with LiCl-induced sickness. Instead, responding differed between groups during SD exposure. Third, if the paired animals did make associations of the nonspecific sensory properties of the infusion with LiCl-induced sickness, these would likely have been extinguished following the completion of devaluation training and thirty days of abstinence during which over 3800 infusions of saline were administered per rat.

One might also consider the possibility that a S^D paradigm might not be sufficient to produce habitual behavior, as goal-directed cigarette-seeking behavior is repeatedly observed in addicted smokers in response to S^D cues [55]–[61]. However, research has shown that resistance to outcome devaluation (e.g. habitual responding) can develop for oral sucrose or food self-administration using similar discrete noncontingent S^D paradigms [62], [63]. Moreover, the present paradigm provided stimuli in addition (i.e. lever, operant chamber) to the S^D that proved sufficient to produce habitual behavior.

The observed goal-directed responding stands in contrast with experimenter-administered psychostimulant experiments leading to habit formation in responding for sucrose [9]–[11]. Since subjects in the current experiment self-administered nearly 600 mg/kg of cocaine over three weeks and the aforementioned study utilizing cocaine involved approximately 400 mg/kg of experimenter-administered cocaine [9], the finding that S^D-induced responding did not become habitual cannot be attributed to insufficient cumulative drug exposure. Instead, one reason that cocaine-seeking remained goal-directed while food-seeking behavior became habitual after drug exposure may be due to fundamental differences between drug-seeking and food-seeking behavior. On the other hand, that habitual responding occurred in the hour prior to S^D presentations and food-seeking behavior can become habitual under similar experimental circumstances following amphetamine-exposure [10], [11] suggest that the expression of habitual behavior is related to presentation of cues that do not strongly correlate with consumption of cocaine or food.

Neural Mechanisms of Relapse

The dorsolateral striatum has been linked with acquiring habitual responding for food reward [24] and has thus been hypothesized to be involved in "habitual drug-seeking" [5]. Macey and colleagues [64] observed decreased glucose metabolism in the dorsolateral striatum after 60 hours of cocaine self-administration (2 hour daily sessions over 30 days). Similarly, a decrease was observed in dorsolateral striatum single neuron firing rates during instrumental movements over 28 hours of water self-administration (2 hour daily sessions over 14 days; [65]). In the water-seeking experiment, animals acquired a habit, as evidenced by maintained operant movements despite prior satiation with water. In the cocaine-seeking experiment [64], although habit formation was not tested, neural activity of dorsolateral striatum neurons could be interpreted as a correlate of habit formation.

A current theory of the neural basis of addiction posits an increasing role of the dorsolateral striatum and a decreasing role of the ventromedial striatum concomitant with a shift from goal-directed to "habitual drug-seeking" [5]. However, ventromedial striatal neurons (especially of the nucleus accumbens (NAcc) core) continue to exhibit robust changes in firing rates during cocaine self-administration or drug-seeking not under the influence of cocaine after many weeks of self administration and abstinence [66], [67]. In contrast, in a variety of paradigms, the vast majority of dorsolateral striatum neurons exhibit decreased neuronal activity [64], [65], [68] or lose their unconditional movement firing characteristics with overtraining [69], [70]. While the dorsolateral striatum is likely to be involved in some aspects of cocaine-seeking behavior [71], involvement of the dorsolateral striatum in cocaine-seeking behavior does not itself constitute evidence that cocaine-seeking behavior is habitual (as argued in [5]). The continued involvement of the NAcc [22], [66], [67], [72], the involvement of other brain regions known to encode the "value" of learned cocaine-associated stimuli [73]–[83], and the present findings suggest that a value-based neural circuitry may be a critical component in mediating SD-induced responding. However, one might speculate that nonvalue based brain regions linked with habitual responding, such as the dorsolateral striatum, may be particularly active during the hour of testing prior to S^D presentations. Nevertheless, drug-seeking behaviors, which are by definition goal-directed and linked with value-based circuitries, and habitual behaviors, which are by definition not goal-directed and linked with nonvalue-based circuitries, are both likely contributors to the process of relapse.

LiCl-Based Aversion Therapies

Although the present results may indirectly support testing the utility of LiCl aversion therapy in reducing cue-induced relapse in cocaine addicted individuals, this was not our intention. While LiCl is known to block the stereotypical behaviors induced by cocaine [84], low dose 24 hour continuous infusion of LiCl does not block self-administration of cocaine [85]. Furthermore, although LiCl produces robust conditioned place aversion [25], LiCl administration does not block conditioned place preference induced by cocaine [29]. While certain types of aversion therapy have been shown to completely eliminate cocaine cravings in the laboratory [86] it is not known if aversion therapy has lasting effects that decrease relapse outside the laboratory [87]. Indeed, craving can be driven by internal cues such as dysphoria [88], [89], which is likely to be induced by aversion therapy. Furthermore, in the present study, pairing LiCl-induced sickness with cocaine eliminated SD-induced responding. but did not eliminate responding altogether. The attenuated level of responding was not decreased enough to prevent self-administration of cocaine on the second day of testing, a testament to the powerful influence of cocaine. Once under the influence of cocaine, addicted individuals nearly always choose cocaine over other reward choices [41], [42] and animals do not cease responding until "satiated" [15].

Conclusion

Habit learning can be pathological, but as a normal process has been described as adaptive [90], allowing for the cognitive elevation of a primary task via subordination of a more common, well-learned behavior. Therefore, it is not unexpected that rats, upon return to the operant environment (and cues weakly correlated with cocaine infusion), should readily return to the task of lever pressing. What is of particular interest is how the SD 1) immediately interrupted habitual responding which preceded its onset and 2) singularly manifested differences in responding consistent with the value of cocaine. These findings support the claim that relapse is a complex behavioral process involving habitual and goal-directed behaviors that are differentially influenced by cues that vary in their correlation with the cocaine infusion. The relative contribution of habit learning versus goal-directed processing in driving relapse remains to be determined and might ultimately guide treatment strategies. Therapies aimed at altering habitual behavior patterns may limit encountering cues even weakly associated with cocaine. Alternatively, the development of therapeutic approaches may be better informed by evidence that the influence of cues signaling a strong relationship with cocaine infusion availability engages goal-directed actions rather than stimulus-response (i.e. habitual) behaviors.

Materials and Methods

Ethics Statement

Protocols were performed in compliance with the Guide for the Care and Use of Laboratory Animals (NIH, Publications 865–23) and were approved by the Institutional Animal Care and Use Committee, Rutgers University.

Subjects and Surgery

Male Long-Evans rats (n = 14; 325–335 g; Charles River, Wilmington, MA) were implanted with a catheter in the right jugular vein. All details of the surgical procedure and post-operative care have been described in detail elsewhere [91]. Following surgery, animals were randomly assigned to one of two groups: paired or unpaired. Animals were administered 200 μL of heparinized-saline infusions every fifteen minutes throughout the experiment, except during training and testing conditions.

Procedure

During the self-administration phase of the experiment, before the beginning of each daily self-administration session, a nonretractable response lever was mounted on a side wall of a standard operant chamber in which the animal lived. Each lever press in the presence of an audible tone (3.5 kHz, 70 dB) produced an intravenous infusion of cocaine (0.355 mg/kg infusion), terminated the tone, and started an intertone interval (3–6 min). If lever pressing did not occur during a 2 min tone presentation period, the tone was terminated, and an intertone interval began. Each cocaine self-administration session lasted until 80 infusions were earned or 6 hours elapsed, whichever occurred first. Self-administration occurred seven days a week daily for three weeks. To facilitate acquisition of self-administration behavior, animals were shaped to lever press in the presence of the S^D for a 0.71 mg/kg dose on the first day of training. During the shaping session, the S^D was continuously sounded until responding occurred, at which time the

S^D was terminated, cocaine was infused, and a thirty second time out period began. Responses during the time out were recorded but had no programmed consequences. Following time out, the continuous S^D was again initiated. After ten cocaine self-infusions and for the remainder of the self-administration phase, rats were trained under the 2 min S^D duration, 3–6 minute time out schedule of reinforcement. Rats were never drug primed. Animals were housed in the self-administration chamber during the self-administration phase of the experiment.

Following three weeks of daily self-administration, animals were transferred and housed in a wire mesh holding cage for the LiCl phase of the experiment. The overall experimental schematic for the LiCl phase is presented in Figure 1 . For the paired group, on day 1 of the three day cycle, each animal was noncontingently infused with cocaine for 1.5hours according to its self-administered pattern of drug intake on the last day of training. Cocaine infusions were immediately followed by infusions of LiCl (18 mg/kg/infusion). Cessation of LiCl administration occurred when cocaine-induced stereotypy as well as locomotion (operationally defined as alternating limb movements) ceased for at least one minute [25]. On days two and three of the three day cycle, paired animals were not administered LiCl or cocaine. For the unpaired group, on day 1 of the three day cycle, LiCl was administered until locomotion ceased for at least one minute. On day two of the three day cycle, unpaired animals were noncontingently infused with cocaine for 1.5 hours according to each animal's self-administered pattern of drug intake on the last day of training. On day three of the three day cycle, unpaired animals were not administered cocaine or LiCl. For paired subjects, four repetitions of the aforementioned cycle occurred. In comparison to unpaired controls, cocaine-infused subjects required more daily LiCl injections to cease locomotion. To equate LiCl exposure for both groups, in addition to four repetitions of the aforementioned cycle occur, unpaired subjects received 1–2 additional cycles of LiCl exposure, with no additional cocaine exposure during the additional cycles.

Note that our outcome devaluation procedure selectively pairs the outcome, cocaine, with LiCl-induced illness in the paired group. Other methods such as LiCl delivery following cocaine self-administration (i.e. punishment) pair LiCl-induced illness with cocaine, instrumental responding, the self-administration chamber, and the prior presentations of the discriminative stimuli, which would generate ambiguous interpretations of testing data. The present outcome devaluation method allows for testing whether 1) stimuli in the environment, which were never paired with LiCl-induced sickness, or 2) the outcome, cocaine, which was paired with LiCl-induced sickness in the paired group, controls responding during abstinent testing. Also, the cocaine self-administration test following cocaine devaluation does not definitively test whether cocaine self-administration is habitual or goal-directed because cocaine is a stimulus which can engender responding on its own [15].

Animals remained in the holding cage for 30 additional days after the last cocaine exposure in the LiCl phase. Subsequently, animals were returned to the self-administration chamber 18 to 72 hours before the test of cue-induced responding (details of test in [22]). On day 1 of testing (cue presentation test), the lever was installed and animals were free to lever press without programmed consequence. The SD tone was not presented to the animal during the first hour of testing. During the remaining five hours of the test, the SD tone was presented for 30 seconds every 3–6 minutes. Responses emitted during tone presentations terminated the tone and infused saline (3.755 s), whereas responses emitted while the tone was off had no programmed consequence. On day 2 of testing, animals were allowed to self-administer cocaine with all parameters identical to training.

Statistical Analysis

All outcome variables in the study, e.g., responses/minute, self-administered mg/kg/day, etc., were analyzed as a function of a set of categorical fixed effect independent variables, e.g., group, week, etc., and their interactions using mixed ANOVAs. SAS PROC GLIMMIX (SAS Institute Inc., 2005) was used to run all analyses. All outcome variables were highly skewed and therefore theorized to be gamma distributed rather than normally distributed. Thus, for all outcome variables a gamma distribution with a log link was specified for the outcome variable in the mixed ANOVA. Outcome variables were collected on multiple occasions from each subject, and thus, subject was specified as a random effects variable for those variables. The final solution for the mixed ANOVA model was estimated using maximum pseudo-likelihood marginal expansion. The degrees of freedom in the model were computed using the containment method. Because the data were not normally distributed, the standard errors were computed using the first order residual empirical estimator, also known as the sandwich estimator. All other default settings in PROC GLIMMIX were maintained. Post-hoc simple effects were computed for any overall significant interactions. Alpha criterion for all tests was 0.05.

Acknowledgements

We thank Linda King, Thomas Grace Sr., Smruti Patel, Dana Silagi, Alyssa Ames, Karina Gotliboyim, Abigail Klein, and Shaili Jha for excellent assistance. We wish to thank three anonymous reviewers for their excellent, scholarly reviews.

Author Contributions

Conceived and designed the experiments: DHR MOW. Performed the experiments: DHR DJB SM. Analyzed the data: DHR DJB APP. Wrote the paper: DHR ATF DJB SM APP MOW.

References

1. Jaffe JH (1990) Drug addiction and drug abuse. In: Gilman AG, Rall TW, Nies AS, Taylor P, editors. Goodman and Gilman's The Pharmacological Basis of Therapeutics. New York: Pergamon. pp. 522–557.
2. Tiffany ST (1990) A cognitive model of drug urges and drug-use behavior: role of automatic and nonautomatic processes. Psychol Rev 97(2): 147–168.
3. Redish AD, Jensen S, Johnson A (2008) A unified framework for addiction: vulnerabilities in the decision process. Behav Brain Sci 31(4): 415–437.
4. Belin D, Everitt BJ (2008) Cocaine seeking habits depend upon dopamine-dependent serial connectivity linking the ventral with the dorsal striatum. Neuron 57(3): 432–441.
5. Everitt BJ, Belin D, Economidou D, Pelloux Y, Dalley JW, et al. (2008) Neural mechanisms underlying the vulnerability to develop compulsive drug-seeking habits and addiction. Philos Trans R Soc Lond B Biol Sci 363(1507): 3125–3135.
6. Fuchs RA, Branham RK, See RE (2006) Different neural substrates mediate cocaine seeking after abstinence versus extinction training: a critical role for the dorsolateral caudate-putamen. J Neurosci 26(13): 3584–3588.
7. Dickinson A (1985) Actions and habits: the development of behavioural autonomy. Philos Trans R Soc London B308: 67–78.
8. Marlatt GA (1985) Relapse prevention: Theoretical rational and overview of the model. In: Marlatt GA, Gordon JR, editors. Relapse prevention. New York: Guilford Press. pp. 250–280.
9. Schoenbaum G, Setlow B (2004) Cocaine makes actions insensitive to outcomes but not extinction: implications for altered orbitofrontal-amygdalar function. Cereb Cortex 15(8): 1162–1169.
10. Nelson A, Killcross S (2006) Amphetamine exposure enhances habit formation. J Neurosci 26(14): 3805–3812.
11. Nordquist RE, Voorn P, de Mooij-van Malsen JG, Joosten RN, Pennartz CM, et al. (2007) Augmented reinforcer value and accelerated habit formation after repeated amphetamine treatment. Eur Neuropsychopharmacol 17(8): 532–540.

12. Miles FJ, Everitt BJ, Dickinson A (2003) Oral cocaine seeking by rats: action or habit? Behav Neurosci 117(5): 927–938.
13. Pickens R, Thompson T (1968) Cocaine-reinforced behavior in rats: effects of reinforcement magnitude and fixed-ratio size. J Pharmacol Exp Ther 161(1): 122–129.
14. Smith BJ, Jones HE, Griffiths RR (2001) Physiological, subjective and reinforcing effects of oral and intravenous cocaine in humans. Psychopharmacology (Berl) 156(4): 435–444.
15. Norman AB, Tsibulsky VL (2006) The compulsion zone: a pharmacological theory of acquired cocaine self-administration. Brain Res 1116(1): 143–152.
16. Peoples LP, Uzwiak AJ, Gee F, West MO (1997) Operant behavior during sessions of intravenous cocaine infusion is necessary and sufficient for phasic firing of single nucleus accumbens neurons. Brain Research 757: 280–284.
17. Epstein DH, Preston KL (2003) The reinstatement model and relapse prevention: a clinical perspective. Psychopharmacology (Berl) 168(1–2): 31–41.
18. O'Brien CP (1976) Experimental analysis of conditioning factors in human narcotic addiction. Pharmacological Reviews 27(4): 533–543.
19. Ehrman RN, Robbins SJ, Childress AR, O'Brien CP (1992) Conditioned responses to cocaine-related stimuli in cocaine abuse patients. Psychopharmacology (Berl) 107(4): 523–529.
20. Smelson DA, Roy A, Roy M, Tershakovec D, Engelhart C, et al. (2001) Electroretinogram and cue-elicited craving in withdrawn cocaine-dependent patients: a replication. Am J Drug Alcohol Abuse 27(2): 391–397.
21. Childress AR, Ehrman RN, Wang Z, Li Y, Sciortino N, et al. (2008) Prelude to passion: limbic activation by "unseen" drug and sexual cues. PLoS One 3(1): e1506.
22. Ghitza UE, Fabbricatore AT, Prokopenko V, Pawlak AP, West MO (2003) Persistent cue-evoked activity of accumbens neurons after prolonged abstinence from self-administered cocaine. J Neurosci 23(19): 7239–7245.
23. Ahmed SH, Koob GF (1998) Transition from moderate to excessive drug intake: change in hedonic set point. Science 282(5387): 298–300.
24. Yin HH, Knowlton BJ, Balleine BW (2004) Lesions of dorsolateral striatum preserve outcome expectancy but disrupt habit formation in instrumental learning. Eur J Neurosci 19(1): 181–189.
25. Mucha RF, van der Kooy D, O'Shaughnessy M, Bucenieks P (1982) Drug reinforcement studied by the use of place conditioning in rat. Brain Res 243(1): 91–105.

26. Balleine BW, Garner C, Gonzalez F, Dickinson A (1995) Motivational control of heterogenous instrumental chains. Journal of Experimental Psychology: Animal Behavior processes 21(3): 203–217.
27. Adams CD (1982) Variations in the sensitivity of instrumental responding to reinforcer devaluation. Quarterly journal of experimental psychology 34B: 77–98.
28. Adams CD, Dickinson A (1981) Instrumental responding following reinforcer devaluation. Quarterly Journal of Experimental Psychology 33B: 109–122.
29. Suzuki T, Shiozaki Y, Masukawa Y, Misawa M, Nagase H (1992) The Role of Mu- and Kappa-Opioid Receptors in Cocaine-Induced Conditioned Place Preference. The Japanese Journal of Pharmacology 58(4): 435–442.
30. Caprioli D, Celentano M, Dubla A, Lucantonio F, Nencini P, et al. (2009) Ambience and drug choice: cocaine- and heroin-taking as a function of environmental context in humans and rats. Biol Psychiatry 65: 893–899.
31. Woolverton WL, English JA (1997) Further analysis of choice between cocaine and food using the unit price model of behavioral economics. Drug Alcohol Depend 49(1): 71–78.
32. Anderson KG, Woolverton WL (2000) Concurrent variable-interval drug self-administration and the generalized matching law: a drug-class comparison. Behav Pharmacol 11(5): 413–420.
33. Nader MA, Woolverton WL (1992) Choice between cocaine and food by rhesus monkeys: effects of conditions of food availability. Behav Pharmacol 3(6): 635–638.
34. Woolverton WL, Anderson KG (2006) Effects of delay to reinforcement on the choice between cocaine and food in rhesus monkeys. Psychopharmacology (Berl) 186(1): 99–106.
35. Woolverton WL, Wang Z (2004) Relationship between injection duration, transporter occupancy and reinforcing strength of cocaine. Eur J Pharmacol 486(3): 251–257.
36. Carroll ME (1985) The role of food deprivation in the maintenance and reinstatement of cocaine-seeking behavior in rats. Drug Alcohol Depend 16(2): 95–109.
37. Woolverton WL, Rowlett JK (1998) Choice maintained by cocaine or food in monkeys: effects of varying probability of reinforcement. Psychopharmacology (Berl) 138(1): 102–106.
38. Lenoir M, Serre F, Cantin L, Ahmed SH (2007) Intense sweetness surpasses cocaine reward. PLoS ONE 2(1): e698.

39. Foltin RW (1999) Food and cocaine self-administration in baboons: effects of alternatives. J Exp Anal Behav 72(2): 215–234.
40. Haney M, Collins ED, Ward AS, Foltin RW, Fischman MW (1999) Effect of a selective dopamine D1 agonist (ABT-431) on smoked cocaine self-administration in humans. Psychopharmacology (Berl) 143(1): 102–110.
41. Donny EC, Bigelow GE, Walsh SL (2003) Choosing to take cocaine in the human laboratory: effects of cocaine dose, inter-choice interval, and magnitude of alternative reinforcement. Drug Alcohol Depend 69(3): 289–301.
42. Donny EC, Bigelow GE, Walsh SL (2004) Assessing the initiation of cocaine self-administration in humans during abstinence: effects of dose, alternative reinforcement, and priming. Psychopharmacology (Berl) 172(3): 316–323.
43. Hart CL, Haney M, Foltin RW, Fischman MW (2000) Alternative reinforcers differentially modify cocaine self-administration by humans. Behav Pharmacol 11(1): 87–91.
44. Hart CL, Haney M, Vosburg SK, Rubin E, Foltin RW (2007) Gabapentin does not reduce smoked cocaine self-administration: employment of a novel self-administration procedure. Behav Pharmacol 18(1): 71–75.
45. Lau-Barraco C, Schmitz JM (2008) Drug preference in cocaine and alcohol dual-dependent patients. Am J Drug Alcohol Abuse 34(2): 211–217.
46. Ward AS, Haney M, Fischman MW, Foltin RW (1997) Binge cocaine self-administration by humans: smoked cocaine. Behav Pharmacol 8(8): 736–744.
47. Ward AS, Haney M, Fischman MW, Foltin RW (1997) Binge cocaine self-administration in humans: intravenous cocaine. Psychopharmacology (Berl) 132(4): 375–381.
48. Foltin RW, Fischman MW (1997) A laboratory model of cocaine withdrawal in humans: intravenous cocaine. Exp Clin Psychopharmacol 5(4): 404–11.
49. Foltin RW, Haney M (2000) Conditioned effects of environmental stimuli paired with smoked cocaine in humans. Psychopharmacology (Berl) 149(1): 24–33.
50. Childress AR, Mozley PD, McElgin W, Fitzgerald J, Reivich M, et al. (1999) Limbic activation during cue-induced cocaine craving. Am J Psychiatry 156(1): 11–18.
51. Wang GJ, Volkow ND, Fowler JS, Cervany P, Hitzemann RJ, et al. (1999) Regional brain metabolic activation during craving elicited by recall of previous drug experiences. Life Sciences 64(9): 775–784.

52. Garavan H, Pankiewicz J, Bloom A, Cho JK, Sperry L, et al. (2000) Cue-induced cocaine craving: neuroanatomical specificity for drug users and drug stimuli. Am J Psychiatry 157(11): 1789–1798.
53. Lynch WJ, Sughondhabirom A, Pittman B, Gueorguieva R, Kalayasiri R, et al. (2006) A paradigm to investigate the regulation of cocaine self-administration in human cocaine users: a randomized trial. Psychopharmacology (Berl) 185(3): 306–314.
54. Donny EC, Bigelow GE, Walsh SL (2006) Comparing the physiological and subjective effects of self-administered vs yoked cocaine in humans. Psychopharmacology (Berl) 186(4): 544–552.
55. Hogarth L, Dickinson A, Duka T (2005) Explicit knowledge of stimulus-outcome contingencies and stimulus control of selective attention and instrumental action in human smoking behaviour. Psychopharmacology (Berl) 177(4): 428–437.
56. Hogarth L, Dickinson A, Hutton SB, Bamborough H, Duka T (2006) Contingency knowledge is necessary for learned motivated behaviour in humans: relevance for addictive behaviour. Addiction 101(8): 1153–1166.
57. Hogarth L, Dickinson A, Hutton SB, Elbers N, Duka T (2006) Drug expectancy is necessary for stimulus control of human attention, instrumental drug-seeking behaviour and subjective pleasure. Psychopharmacology (Berl) 185(4): 495–504.
58. Hogarth L, Duka T (2006) Human nicotine conditioning requires explicit contingency knowledge: is addictive behaviour cognitively mediated? Psychopharmacology (Berl) 184(3–4): 553–566.
59. Hogarth L, Dickinson A, Wright A, Kouvaraki M, Duka T (2007) The role of drug expectancy in the control of human drug seeking. J Exp Psychol Anim Behav Process 33(4): 484–396.
60. Hogarth L, Dickinson A, Janowski M, Nikitina A, Duka T (2008) The role of attentional bias in mediating human drug-seeking behaviour. Psychopharmacology (Berl) 201(1): 29–41.
61. Hogarth L, Dickinson A, Duka T (2009) Detection versus sustained attention to drug cues have dissociable roles in mediating drug seeking behavior. Exp Clin Psychopharmacol 17(1): 21–30.
62. Wilson CL, Sherman JE, Holman EW (1981) Aversion to the reinforcer differentially affects conditioned reinforcement and instrumental responding. J Exp Psychol Anim Behav Process 7(2): 165–174.

63. Callu D, Puget S, Faure A, Guegan M, El Massioui (2007) Habit learning dissociation in rats with lesions to the vermis and the interpositus of the cerebellum. Neurobiology of Disease 27: 228–237.
64. Macey DJ, Rice WN, Freedland CS, Whitlow CT, Porrino LJ (2004) Patterns of functional activity associated with cocaine self-administration in the rat change over time. Psychopharmacology (Berl) 172(4): 384–92.
65. Tang C, Pawlak AP, Prokopenko V, West MO (2007) Dose- and rate-dependent effects of cocaine on striatal firing related to licking. J Pharmacol Exp Ther 324(2): 701–713.
66. Ghitza UE, Fabbricatore AT, Prokopenko VF, West MO (2004) Differences between accumbens core and shell neurons exhibiting phasic firing patterns related to drug-seeking behavior during a discriminative stimulus task. J Neurophysiol 92(3): 1608–1614.
67. Hollander JA, Carelli RM (2005) Abstinence from cocaine self-administration heightens neural encoding of goal-directed behaviors in the accumbens. Neuropsychopharmacology 30: 1464–1474.
68. Porrino LJ, Lyons D, Smith HR, Daunais JB, Nader MA (2004) Cocaine self-administration produces a progressive involvement of limbic, association, and sensorimotor striatal domains. J Neurosci 24(14): 3553–3562.
69. Carelli RM, Wolske M, West MO (1997) Loss of lever press-related firing of rat striatal forelimb neurons after repeated sessions in a lever pressing task. J Neurosci 17(5): 1804–1814.
70. Tang C, Pawlak AP, Prokopenko V, West MO (2007) Changes in activity of the striatum during formation of a motor habit. Eur J Neurosci 25: 1212–1227.
71. Fuchs RA, Branham RK, See RE (2006) Different neural substrates mediate cocaine seeking after abstinence versus extinction training: a critical role for the dorsolateral caudate-putamen. J Neurosci 26(13): 3584–3588.
72. Hollander JA, Carelli RM (2007) Cocaine-associated stimuli increase cocaine seeking and activate accumbens core neurons after abstinence. J Neurosci, 3535–3539.
73. Ciccocioppo R, Sanna PP, Weiss F (2001) Cocaine-predictive stimulus induces drug-seeking behavior and neural activation in limbic brain regions after multiple months of abstinence: reversal by D(1) antagonists. Proc Natl Acad Sci USA 98(4): 1976–1981.
74. Kruzich PJ, See RE (2001) Differential contributions of the basolateral and central amygdala in the acquisition and expression of conditioned relapse to cocaine-seeking behavior. J Neurosci 21(14): 1–5.

75. Kantak KM, Black Y, Valencia E, Green-Jordan K, Eichenbaum HB (2002) Dissociable effects of lidocaine inactivation of the rostral and caudal basolateral amygdala on the maintenance and reinstatement of cocaine-seeking behavior in rats. J Neurosci 22(3): 1126–1136.

76. Hayes RJ, Vorel SR, Spector J, Liu X, Gardner EL (2003) Electrical and chemical stimulation of the basolateral complex of the amygdala reinstates cocaine-seeking behavior in the rat. Psychopharmacology (Berl) 168(1–2): 75–83.

77. Yun IA, Fields HL (2003) Basolateral amygdala lesions impair both cue- and cocaine-induced reinstatement in animals trained on a discriminative stimulus task. Neuroscience 121(3): 747–757.

78. Di Ciano P, Everitt BJ (2004) Direct interactions between the basolateral amygdala and nucleus accumbens core underlie cocaine-seeking behavior by rats. J Neurosci 24(32): 7167–7173.

79. Fuchs RA, Evans KA, Ledford CC, Parker MP, Case JM, et al. (2005) The role of the dorsomedial prefrontal cortex, basolateral amygdala, and dorsal hippocampus in contextual reinstatement of cocaine seeking in rats. Neuropsychopharmacology 30(2): 296–309.

80. See RE (2005) Neural substrates of cocaine-cue associations that trigger relapse. Eur J Pharmacol 526(1–3): 140–146.

81. Berglind WJ, Case JM, Parker MP, Fuchs RA, See RE (2006) Dopamine D1 or D2 receptor antagonism within the basolateral amygdala differentially alters the acquisition of cocaine-cue associations necessary for cue-induced reinstatement of cocaine-seeking. Neuroscience 137(2): 699–706.

82. Peters J, Vallone J, Laurendi K, Kalivas PW (2008) Opposing roles for the ventral prefrontal cortex and the basolateral amygdala on the spontaneous recovery of cocaine-seeking in rats. Psychopharmacology (Berl) 197(2): 319–326.

83. Stalnaker TA, Roesch MR, Franz TM, Calu DJ, Singh T, et al. (2007) Cocaine-induced decision-making deficits are mediated by miscoding in basolateral amygdala. Nat Neurosci 10(8): 949–951.

84. Flemenbaum A (1977) Antagonism of behavioral effects of cocaine by lithium. Pharmacol Biochem Behav 7(1): 83–85.

85. Woolverton WL, Balster RL (1979) The effects of lithium on choice between cocaine and food in the rhesus monkey. Commun Psychopharmacol 3(5): 309–318.

86. Bordnick PS, Elkins RL, Orr TL, Walters P, Thyer BA (2004) Evaluating the relative effectiveness of three aversion therapies designed to reduce craving among cocaine abusers. Behavioral Interventions 19: 1–24.

87. McLellan AT, Childress AR (1985) Aversive therapies for substance abuse: do they work? J Subst Abuse Treat 2(3): 187–191.

88. Wallace BC (1989) Psychological and environmental determinants of relapse in crack cocaine smokers. J Subst Abuse Treat 6(2): 95–106.

89. Gawin FH, Khalsa-Denison ME (1996) Is craving mood-driven or self-propelled? Sensitization and "street" stimulant addiction. NIDA Res Monogr 163: 224–250.

90. Balleine BW, Liljeholm M, Ostlund SB (2009) The integrative function of the basal ganglia in instrumental conditioning. Behav Brain Res 199(1): 43–52.

91. Peoples LP, West MO (1996) Phasic firing of single neurons in the rat nucleus accumbens correlated with the timing of intravenous cocaine self-administration. J Neurosci 16(10): 3459–3473.

Conflict and User Involvement in Drug Misuse Treatment Decision-Making: A Qualitative Study

Jan Fischer, Joanne Neale, Michael Bloor and Nicholas Jenkins

ABSTRACT

Background

This paper examines client/staff conflict and user involvement in drug misuse treatment decision-making.

Methods

Seventy-nine in-depth interviews were conducted with new treatment clients in two residential and two community drug treatment agencies. Fifty-nine of these clients were interviewed again after twelve weeks. Twenty-seven interviews were also conducted with staff, who were the keyworkers for the interviewed clients.

Results

Drug users did not expect, desire or prepare for conflict at treatment entry. They reported few actual conflicts within the treatment setting, but routinely discussed latent conflicts–that is, negative experiences and problematic aspects of current or previous treatment that could potentially escalate into overt disputes. Conflict resulted in a number of possible outcomes, including the premature termination of treatment; staff deciding on the appropriate outcome; the client appealing to the governance structure of the agency; brokered compromise; and staff skilfully eliciting client consent for staff decisions.

Conclusion

Although the implementation of user involvement in drug treatment decision-making has the potential to trigger high levels of staff-client conflict, latent conflict is more common than overt conflict and not all conflict is negative. Drug users generally want to be co-operative at treatment entry and often adopt non-confrontational forms of covert resistance to decisions about which they disagree. Staff sometimes deploy user involvement as a strategy for managing conflict and soliciting client compliance to treatment protocols. Suggestions for minimising and avoiding harmful conflict in treatment settings are given.

Background

In recent years, user involvement has become an important concept in health and social care policy and practice. Its origins and development have variously been related to the anti-psychiatry movement, the rise of consumerism, the emergence of self advocacy and pressure groups, the growth of community action, New Right policies, and the increase in public willingness to question expert knowledge in late modern society [1-5]. Today user involvement is commonplace in health fields as diverse as cancer treatment, mental health, learning disabilities and maternity services, with the research literature indicating that it improves service provision (e.g. [6]), empowers individuals (e.g. [7]) and is a democratic right and an ethical requirement (e.g. [8]).

Although user involvement has been relatively slow to develop within the UK drug treatment field, steady progress has been made and a strong user involvement movement now exists (as exemplified by Narcotics Anonymous, The Alliance, and Mainliners) [9]. Drug user representation at the level of national policy making remains poor, but local Drug (and Alcohol) Action Teams are involving users through user groups, user involvement co-ordinators, and user consultations. In addition, regional users' forums have been established, and the National

Treatment Agency (NTA) in England has written user involvement clearly into recent policy documents, stating for example that 'Service users should be involved in all key aspects of decision-making in relation to their care' [10]. This paper focuses on the involvement of drug users in making decisions about their own treatment, a practice also referred to as client-centred treatment (e.g. [11-13]).

Implementing user involvement within the substance misuse field is, however, unlikely to be straightforward. Over the years, many authors have portrayed drug users as impatient, manipulative and aggressive, and identified hostility and anger as typical drug user characteristics [14-19]. Indeed, De Leon has argued that drug users 'often display an extreme sense of entitlement and exaggerated reactions to perceived unfairness, a need for immediate gratification in the form of instant answers, resistance through arguments, and a tendency to manipulate authority figures' [20]. Involving such individuals–particularly in treatment decision-making–seems prone to difficulty, especially if they breach treatment protocols and misuse treatment facilities [21]. Furthermore, a strong blame culture within the drug treatment field means that drug users are often seen as undeserving and not consulted despite policy statements [22].

In some treatment settings, particularly residential settings, staff may also actively provoke (and subsequently manage) conflict with drug users as part of the therapeutic process. Known as 'reality confrontation', the intention is to contain dysfunctional behaviours and enable clients to discover that they can tolerate uncomfortable emotional states and learn appropriate responses to difficult situations [23-26]. Such activity involves the repetitive depiction to clients of their behaviour as unacceptable, alongside the depiction of the community as a locale where other less pathogenic ways of behaving could be experimented with and adopted [27]. In addition, therapeutic communities for drug users are often (following the original 'Synanon' model [28]) hierarchically organised with a highly structured programme of activities and elaborate systems of rules. The expectation is that residents will fall foul of the rules and perform inadequately in the programme, but these failures offer valuable opportunities for therapeutic work [29].

Taking all the above factors into consideration, it seems reasonable to assume that involving drug users in making treatment decisions will create tensions, disagreements and even outright conflict between clients and staff. However, a broader reading of the sociological literature cautions against any simplistic assumptions. General writings on the doctor-patient relationship have long suggested that overt conflict in the healthcare setting is rare [30-32] and work on therapeutic communities for those with mental health problems indicates that dissenting clients will often use concealment and other forms of resistance in their interaction with staff, rather than resorting to open conflict [33]. As Goffman notes, residents of institutional settings will frequently avoid open conflict with staff by making secondary adjustments, such as working the system, taking back some small measure of control over their immediate environment and feigning to accept the negative views that staff have of them [34].

Certainly, user involvement has become a popular concept in substance misuse treatment policy and practice. Nonetheless, the existing literature does not provide a clear picture of whether and, if so, how involving drug users in treatment decision-making might result in conflict, how individuals might react to that conflict, and what the impact of any conflict might be. In this paper, we therefore seek to develop our understanding of user involvement by examining the extent, causes, responses to, and outcomes of conflicts occurring between the clients and staff of drug treatment services. We also consider whether conflict between clients and staff is always negative and how any negative forms might be prevented or reduced.

Methods

The paper draws on data from in-depth semi-structured interviews conducted with 79 drug users and 27 treatment agency staff. The 79 drug users were recruited from two community drug treatment services (n = 19 and n = 20) and two residential rehabilitation agencies (n = 20 each). They included 53 men and 26 women, who were all interviewed within ten days of beginning a new treatment episode and who were all seeking treatment for their use of illicit drugs (predominantly opiates and/or stimulants). Their prior treatment histories were very varied (from non-existent to lengthy), with the clients of the residential services tending to report the most previous treatment episodes.

Contact details for all drug users were recorded at first interview and efforts were made to re-contact them after twelve weeks. In total, 59 were successfully re-interviewed (n = 14 and n = 15 from the two community services; n = 14 and n = 16 from the two residential services). Those who were not re-interviewed could not be traced at any known address, repeatedly failed to keep appointments with the interviewer or refused to participate in a second interview. Of the 29 community treatment clients who were re-interviewed, only 3 had left treatment at follow-up. In contrast, 16 of the 30 re-interviewed residential treatment clients had left treatment by their second interview. Most of these 16 individuals said that their treatment had ended for negative reasons, rather than because they had been ready to move on.

The 27 treatment agency staff who were interviewed were all nursing or care staff who had been selected to participate in the research because they were the designated key worker of one or more of the clients interviewed. Ethical approval for all interviews was granted by the Thames Valley Multi-centre Research Ethics Committee (approval number: 05/MRE12/48) and research was compliant with the Helsinki Declaration. All drug users and staff gave written informed consent prior to each interview and, during their first interview, all drug users consented to being re-contacted after twelve weeks for a follow-up interview.

Towards the end of each interview, drug users and staff were asked to engage with a developmental vignette, a research technique previously described in the drug misuse literature by Hughes [35]. The vignette told the story of a fictional drug user's treatment career and incorporated a number of scenarios with potential for conflict relating to treatment decision-making. These included the protagonist being offered a relatively limited methadone prescription, being forced to participate in group work, and being asked to own up to using drugs within a residential setting. Interviewees were asked to state how they thought the character would respond to each of these events. The vignette and more details on the research methods are reported elsewhere [9].

Both the semi-structured interviews and the vignette data were transcribed and coded with the aid of the software package MAXqda2. Codes from the interviews relating to past treatment experiences, experiences of the current treatment episode, and future expectations regarding the current treatment episode were exported into Microsoft Word files and systematically searched for evidence of conflicts relating to treatment decisions. These instances were then analysed thematically using Framework [36]. Responses to the vignette stages were similarly analysed and also used to appraise the reliability of the interview data. Although the client interviews were drawn from two time points, a longitudinal analysis is not presented here as we wanted to incorporate data from all of the clients' treatment experiences (previous and current). This was for two reasons: first, we wanted to maximise the amount of analysable data; second, we wanted to counter the likelihood that clients might give socially desirable responses when discussing their current treatment episode.

Preliminary findings indicated that 'conflict' was a very broad term that encompassed relatively trivial incompatibilities of opinion through to major disputes that resulted in treatment breakdown. After team discussions, we therefore decided to employ the terms 'actual conflict' and 'latent conflict' to further the analyses. We used the term 'actual conflict' to refer to those situations which had already resulted in open hostilities or clearly expressed differences of opinion. In contrast, we used 'latent conflict' to refer to situations that had not progressed into open disagreements but had the clear potential to develop into serious arguments and to describe signs of disagreement hidden below a veneer of consensus.

Results

The Extent of Conflict

During their first interview, most clients reported that they were not anticipating any conflicts in their current treatment episode. Furthermore, there was no indication that drug users from either the residential or community settings expected that staff would deliberately initiate conflict as part of the therapeutic/treatment process. Despite this, clients from the two residential units were more likely than clients from the two community services to report that future disputes might occur. These differences between service settings seemed to relate to two factors. Firstly, the residential clients–having generally had more previous treatments than the community clients–were more aware of the kinds of conflicts that could arise, particularly as they had often been referred to residential treatment after failing to stabilise in a community setting. Secondly, residential treatment is a more intense experience than community treatment and so more prone to generating stresses and strained interpersonal relations.

The follow-up interviews generally confirmed the clients' expectations, with fewer reports of conflicts in the community agencies than in the residential agencies. These findings were partially mirrored in the staff interviews. Although expectations of conflict varied substantially between individual staff members, residential workers appeared to expect a higher frequency of conflicts than those employed in community settings.

Given that most clients did not anticipate problems with staff at their first interviews, it was not surprising to find that they had not pre-prepared any strategies for dealing with disputes or disagreements. This is contrary to what one might have expected from the image of the impatient and manipulative drug user, with an extreme sense of entitlement, described in some of the literature discussed earlier. Instead, many treatment clients spoke of approaching their first treatment contact with openness and honesty, in the hope that all would go well and they would receive much-needed help and support. For example, the following statement was made by a female community client who had accepted her mother's advice regarding the importance of being honest with services:

> *One thing my mum said is just to make sure I'm honest, do you know what I mean? There's no point in me coming down to somebody unless I'm being honest, 'cos otherwise I'll not get the help I need, do you know what I mean? So there's no point in me trying to hide anything. [community client, first interview]*

Additionally, when clients were asked what they expected their first meeting with the treatment agency staff would be like, most said that they had never thought about the issue. As this male community agency client explained:

I have never really thought about it. I just take it as it comes. I shall be as polite as I can and talk as normal as I can to them and hope they respect me. [community client, first interview]

Those who had given the matter of the first treatment encounter any consideration tended to report that they wanted, and were expecting, to be guided by staff rather than to take the lead themselves. This was because they perceived staff as being the experts who would know the best course of action to take and be able to allay their fears and anxieties. Moreover, staff who were themselves ex-drug users were deemed to have the greatest knowledge, understanding and credibility. Similar findings emerged from the vignette. When interviewees were asked whether the therapist in the fictional story was right or wrong to insist that the protagonist attend a group session against his will, most clients thought that attendance should be compulsory. Frequently, they justified this by reference to the therapists' expertise or a belief that group work must be beneficial if it was a formal part of the treatment programme:

I was apprehensive at the very beginning [about group sessions] but once you do start attending them you do realise where the therapists are coming from. Most therapists are actually ex-users and they know what they are talking about. So, I believe that they are right. [residential client, first interview]

Although the employment of ex-users varied substantially between our four treatment agencies (with one residential agency employing almost exclusively ex-users and one community agency employing none), staff interviews often confirmed that ex-users as staff enjoyed greater credibility than staff who were not ex-users. Furthermore, despite a general acceptance of user involvement, staff seemed to agree with clients in favouring a staff-led service:

They [clients] certainly have a say in it in the sense that within a week you have to agree a detailed treatment plan with the patient and with the head of treatment here. If the patient thinks it is wrong, they can certainly say so. The patient can't really dictate more than that. I mean, they can't come in here with their own ideas about what they can and can't do, but they do certainly have a say. [residential staff member]

This symmetry of client and staff viewpoints was less evident at the second client interviews. Although the majority of clients still believed that clients should defer to staff expertise, a minority now felt that more user participation in treatment decisions would be beneficial:

I think it should all be decided by them [staff]. Because, at the end of the day, at the start of treatment you're not really capable of deciding what is best for yourself. [residential client, first interview]

I just think, if the people [clients] had more say, they'd feel more comfortable in the things, especially for the newer people. [same client, follow-up interview]

About half of those who were re-interviewed reported that they had experienced conflict (actual or latent) within their ongoing, or by then terminated, treatment. Some clients referred to isolated, low-level latent disputes, but a few stated that they had disagreed with staff in a more or less open way on a number of occasions. The majority of clients who had left treatment between their first and second interview had experienced conflict of some kind and this conflict had often related to their treatment ending. Amongst those clients who remained in treatment at their second interview, initial enthusiasm for open communication had sometimes diminished because of a conflict experience (the causes of which are examined below).

The Causes of Conflict

In the early stages of treatment, many of the causes of conflict discussed by clients related to aspects of the treatment programme that angered or irritated them because they did not understand them or find them helpful. In residential rehabilitation settings, clients commonly complained about 'silly' or 'petty' rules and particular programme procedures. These included restrictions on music, television and phone calls; not being allowed to consume food brought onto the premises; and not being allowed to go to one's room alone. Examples of procedures that irritated clients were unnecessary domestic chores; being constantly chaperoned; and having to participate in group work. This interviewee, for example, describes how some of these rules had made her very angry:

I didn't know I wasn't allowed to bring anything up, a hi-fi, whatever...And I went out and spent money and bought cranberry juice, things that I drink. [Staff member] took half my things off me. I was angry about that...And I was feeling really lonely, really hurt. I just wanted to go to my room and cry. And then I was told you have got to get permission to do that. [residential client, first interview]

When clients felt that they did not understand the reasoning behind rules and procedures, low level discontent and irritation (latent conflict) sometimes escalated into actual conflict. Conversely, clients' annoyance at rules and procedures very often subsided or disappeared if the reasons for such arrangements were clearly explained to them. Thus, in the residential services, frustration at not being allowed to go to one's own room could, with explanation, be reconstructed as positive encouragement to participate in the community. Equally, pointless cleaning duties could be reinterpreted as keeping physically and mentally active. This male client described how he had initially thought that not allowing new residents to listen to music was a 'stupid rule.' Nonetheless, he had come to view it as a 'good idea' after someone had clarified to him the reasoning behind it:

> *I started listening to dance music and I would think of all the good times I had had with dance music on valium and smoking a joint. So for the first six weeks there's no music or television. It's just really focusing on you, because these things make you isolated if you sit and listen to music, and really you should be in the community and talking to people and focusing on you, you know what I mean? So it is a good idea. [residential client, first interview]*

Even though rules, and particularly the highly structured nature of residential treatment, remained a source of discontent for clients at follow-up, those who remained in treatment mostly came to accept that there was no point in arguing about the content of treatment rules as these would never be changed. Instead, client unhappiness seemed to shift to the way that rules were enforced. Clients–particularly residential clients who had themselves been discharged from treatment by staff because of a rule violation–maintained that some staff were too strict or too inflexible in their enforcement of rules. Meanwhile, some community service clients complained that appointment systems were overly strict or that they had been given stricter treatment regimes than others (for example, they had had to consume their medication under supervision or had been subject to frequent urine testing). The strongest resentments, however, seemed to arise when staff imposed different penalties on different clients for the same or similar rule violations:

> *Just the way the place is run, but it is annoying and it just causes resentment. It is like when people go off project drinking. Oh, you go off project drinking, come back, you either come back razzled or you have got this problem going on: 'Okay, we will slap you and don't do it again.' Somebody else does it and it is like: 'Sanction. Seven days notice to quit.' Do you know what I mean? ... I personally [think] it should be one rule for all. [residential client, follow-up]*

Another common source of latent, and some actual, conflict reported by clients was negative staff attitudes and negative staff behaviour. Negative attitudes included staff being uninterested, unsympathetic or looking down on clients. Negative behaviours included staff being remote and uninvolved, not listening to clients, and failing to act–particularly failing to do things that they had promised the client they would do. At the follow-up interviews, only a handful of clients reported that staff had deliberately provoked confrontation and most did not see this as a positive therapeutic measure. On the contrary, staff negativity was overwhelmingly perceived by clients as unhelpful and unconstructive. Indeed, it was occasionally cited as a reason why they had dropped out of treatment.

For a very small number of drug users, actual conflict had occurred when they had wanted a particular treatment but been refused it. This might have happened if a client had requested a treatment which the agency did not provide, or for which the agency had considered them unsuitable, or if the desired treatment had been particularly expensive or difficult to obtain (such as a residential treatment with limited places). For example, this female client had for a time been refused a referral into residential treatment because she was not considered sufficiently stable:

> *It really sent me off my head, because I was like 'how can you not be stable enough for rehab?' And that was her words to me. But she didn't prescribe anything, do you know what I mean? And I was like 'I'm crying out for help here, why are you not helping me?' [residential client, first interview]*

More commonly, latent conflict was evident when drug users anticipated that they would be given a treatment that they particularly disliked, especially methadone. Overt conflict was usually avoided in these situations because the disliked treatment was never actually offered, the client had simply declined it when offered, or the client had conceded and accepted it without complaint. Latent conflict was similarly apparent in the vignette when respondents were asked how the protagonist would feel about being offered only a relatively limited methadone dosage. Many clients reported that this would likely be a source of tension, but often qualified their responses by acknowledging that other drug users would not necessarily be unhappy (or that they would see the offer as a starting dose rather than all they could expect to receive).

Those in residential services additionally stated that the first few days of a residential programme could be particularly prone to arguments, disputes and conflict more generally. This was because clients were often stressed and irritable if they were feeling disorientated, vulnerable, and lonely. Moreover, this was likely to be exacerbated if individuals were undergoing a period of rapid detoxification, since the withdrawal process frayed tempers and made individuals behave unreasonably. Indeed, there was a general feeling that residential clients could not be expected–by staff or others–to be reasonable at such a testing time. Reflecting this, some clients reported that the demands of their detoxification had contributed to them dropping out of treatment prematurely and some called for more or better medication to address the severity of their withdrawal symptoms:

I think there should be medication. They knew that, I think, but four weeks I was up like a baby; stayin' up every night, every night mostly. That's why I got sick o' it. If I'd maybe got a couple o' nights sleep I might have stayed. [residential client, follow-up interview after leaving treatment]

Less frequently, drug users acknowledged that their own bad behaviour and/or negative states of mind were causes of conflict in treatment. Some interviewees thought that their own 'attitude problems' and personal inclinations to 'argue back' at authority were likely to be the source of conflict in their current treatment. Others reported that they had previously dropped out of treatment because they had not really been 'ready' to address their problem. As this client remarked:

It's been my fault, you know...Basically because it comes down to me, you know. I've either relapsed or something...I've had to leave the projects, so usually been down to myself. [community client, first interview]

Staff also identified lack of compliance, rule violations and inappropriate client behaviour as causes of conflict. Indeed, some staff members felt that clients were not truthful about many aspects of their lives and some stated that they would challenge clients who gave contradictory accounts of their treatment progress or who appeared to tell untruths. Despite this, the most common causes of conflict identified by staff were clients' unrealistic expectations about treatment. These included clients expecting medications that were not on offer to them, or that were only offered at low dosage, and not appreciating how long it would take between first seeking treatment and achieving relative stability:

One of the typical sources of conflict is about how soon they think things happen. [community staff member]

Finally, some staff and clients pointed out that conflict inevitably occurs in treatment settings because it is human nature to disagree and argue, especially in residential units where large numbers of people live together. In these cases, our respondents did not distinguish between conflicts that occurred between staff and clients, conflicts that occurred between clients and other clients, and even conflicts within the staff team. Such statements provide a useful reminder that it can never be possible to prevent all conflict in the treatment setting.

Drug Users' Strategies for Dealing with Conflict

As indicated previously, most clients had not prepared a strategy for dealing with any conflict occurring in their current treatment episode. However, when asked what they would do if a serious disagreement arose, most reported that they would try to resolve it by communicating with staff. In this regard, they usually stated that they would explain their position or their side of the story so that staff would better understand things from their point of view. Others stated that they would apologise if they had behaved badly or attempt to talk their way out of a difficult situation. Significantly, many emphasised that they would try to reach a compromise with staff, a finding which again runs counter to the notion of the uncooperative and demanding client. When asked how he would resolve any conflict that might occur in his current treatment, this client responded:

> *Talking. I talk, me. I resolve my problems by talking. [community client, first interview]*

Other clients maintained that they would use avoidance tactics as a means of dealing with potential sources of conflict. This might be avoiding a particular staff member or another service user, refusing a particular element of a treatment programme (such as methadone), or even leaving treatment altogether. This residential client described how he planned to avoid trouble in his current treatment episode by keeping a low profile:

> *I'm having my ups and downs myself at the moment. But I think, as I say, the people seem alright. So I'll just keep my head down and get on with my own thing and I should be alright. [residential client, first interview]*

Staff equally identified negotiation–albeit usually within a framework of rules–as a preferred method for resolving conflicts. Thus, staff described how formal rules and procedures could be used to discipline noncompliant clients and so direct the client's behaviour. Whilst both residential and community staff members felt that some rules were not open to negotiation, the residential settings had the most formalised and elaborate structures. These included written rules about acceptable and unacceptable behaviour and established procedures for taking interpersonal disputes to the wider house community for discussion. Clients of the two residential agencies also reported that they would actively use the formalised structures established in those services to address any actual or potential conflict. Indeed, even clients who had only been in the house for one or two days often had a very good grasp of how the house system could be deployed in this way. As one residential client commented when asked how he might deal with any future conflict:

Use the system. Use the tools and diplomatically. [residential client, first interview]

In contrast, community clients were often less aware of–or less willing to use–formal procedures for addressing grievances, such as the complaints procedures offered by the National Health Service. During follow-up interviews, a number of residential clients identified group work as a useful mechanism for dealing with conflicts in a constructive way. Others felt that group work was at times intimidating and frustrating, and some noted that acceptance of group work was a gradual process as fears were substituted by appreciation of peer advice:

The therapeutic groups are good. You've got people in your group to help you. They see you as you are and it's like they're telling you what's wrong. Me myself was kinda putting barriers up: 'That's not me!' and 'Don't you dare say that about me!' you know? But, after a wee while you kinda sit and you look at it, what they see, because they're actually seeing ya. [residential client, follow-up interview]

Responses to the vignette, meanwhile, indicated that when disputes occurred between staff and clients, many clients believed that they had little option but to concede to staff if they wanted to stay in treatment. In addition, it appeared that honesty could play an important role in dealing with potential conflict. For example, the majority of clients thought that the fictional treatment character in the vignette would be more likely to own up to, than to lie about, using heroin whilst in residential rehabilitation. This was because they felt that the main factors motivating him would be to stay in treatment, do what was morally right, and not to let his house mates down. Showing honesty–rather than arguing or lying–was considered the best ways of proving commitment to the programme, learning from mistakes, and ultimately securing a better treatment outcome. In stark contrast, most staff members expected that the fictional drug user would keep his rule violations secret because he would be afraid of loosing his place in the programme.

Despite their first interview statements about how they would respond to conflict and disagreements by communicating and negotiating, the follow-up interviews showed that users had often not succeeded in addressing problems in a calm, open and diplomatic way. Instead, clients often simply kept quiet about their problems or side-stepped conflict by making 'secondary adjustments', such as purchasing drugs illicitly from the street if they could not obtain them on prescription. Thus, latent conflict never developed into open dispute and client acquiescence could mask hidden dissent:

I mean she is the doctor at the end of the day and if she don't think it right to give them [Valium] to me....So maybe she thinks that I...but if she doesn't give them to me and I go and buy them off the streets anyway so... [community client, second interview]

Other conflicts–albeit rarely–resulted in shouting matches and verbal abuse. An example is provided by this client, who later regretted his response to an argument in which he had been told to sit on a bench (a technique used within the residential agency to calm clients and address disobedience):

I had something put to me. I didn't agree with it and I got told to take the bench and I refused to take the bench and told them to stick it where the sun doesn't shine, basically.... It was out of order; it was very wrong doing that. [residential client, follow-up interview]

Conflict Outcomes

When clients reported on conflicts arising in both their current treatment episode and any episodes of treatment they had had in the past, a number of outcomes seemed possible. One of the most common was the termination of treatment, arising either because the client had left treatment of his or her own accord or because staff had decided to cease providing treatment to them. The kinds of conflict that resulted in clients leaving treatment of their own accord were likely to be latent, relating to such factors as clients disliking the service or aspects of it, feeling that staff had a negative attitude towards them, or perceiving that the treatment was not helping them sufficiently. The kinds of conflict that resulted in staff deciding to end treatment tended to be actual conflicts, resulting from the client being caught tampering with urine samples, using drugs, or selling drugs. For example, this client recalled how he had been discharged from a residential service for taking drugs brought in by another client:

I basically said to them that the whole reason I was there was because I couldn't just say no. I couldn't resist temptation and if a drug was in front of me, I was going to use and that was the whole reason I got asked to leave. [residential client, follow-up interview after discharge]

A further outcome was that staff simply decided on a response, position or course of action and refused to compromise with the client. Thereafter, the issue was not discussed again. In such cases, the staff member would tend to rely on agency rules or policy to uphold their position and the client would defer to them because of their desire for treatment or because they appreciated that they could not win an argument with a professional. In practice, however, the conflict was never actually resolved and so it effectively remained a latent dispute:

Well, at the end of the day, they are professional. They always have the last say. There was nothing that kinda blew up, nothing major, just a couple of disagreements. [residential client, first interview]

For residential clients in particular, conflict could be addressed by appealing to the governance structure of the residential agency. This meant that conflicts, rule violations or other issues that might disrupt treatment were submitted to a group for discussion or that the agency had a rule to deal with particular transgressions, although staff often retained the role of the final decision-maker in both instances. A number of clients seemed to value the process of tackling concerns in a group environment, and some found it therapeutic to address their disputatiousness with others–that is, to be subject to the 'reality confrontation' described earlier:

[Recently] I've been quite angry and 'don't you tell me what to do.'..I am 27 and I've got to stop acting so young....I do need my arse kicked basically, I do. [residential client, second interview]

However, others perceived therapeutic groups as intimidating, as this client explained:

Basically it's jist....well what happens in them [groups] is people get put up for concern like whether it be health or if they're doin' somethin' wrong so that it's voiced in front of like the whole community, so that we're aware and we can help that person. Athough sometimes it disnae feel like that. Ye think they've got the firing guns out at ye. [residential client, follow-up interview]

Another less frequent outcome of conflict was that clients managed to broker a compromise with staff. In these situations, clients generally reported being very happy with the settlement reached and seemed to feel that it had genuinely resolved the conflict. Several staff members also indicated that they preferred a negotiated outcome over any other. In addition, a very small number of clients reported that they had managed to resolve a conflict entirely in their own favour. This had, however, only occurred in situations where the underlying problem was a misunderstanding relating to a technical issue (such as an incorrect prescription) rather than a substantive difference of opinion.

The very limited degree to which clients were able to resolve disputes in their favour is mirrored by staff who reported that skilful staff work involved eliciting client consent for staff decisions:

> *It's that informed choice, um, having to have enough knowledge of what they're actually wanting to talk about and be able to show them the pros and cons of each....and then make them think again about what will work for them. And say: 'Well, you've told me you've done this in the past and that in the past and it hasn't worked and I'm offering this and that and the other....so you need to make that choice. But I would think that the best way from what you're telling me is to go this way, rather than that way. [community staff member]*

For some staff members, it was evident that conflicts could serve as a very useful occasion for therapeutic work. This was because they provided the staff member with opportunities for showing the client Cooley's 'looking glass self' [37]. That is, the staff member could reflect back to the client the image that staff and other clients had of them and their behaviour. Equally, conflicts allowed staff the chance to convert the client to–and get them to agree with–the staff's view of the client's difficulties and needs.

Conclusion

Our study is not without limitations. For example, we did not collect any ethnographic or observational data. This limits our ability to adjudicate between discrepant client and staff accounts of conflict. In addition, our second interviews took place only three months after treatment had started. In some residential agencies, a client is barely seen as settled-in after this time period and some confrontational aspects of therapy may only commence at a later stage. Also, we were not able to re-interview all of the clients. Those not re-interviewed were often no longer in contact with our treatment services and so it is possible that they were discontented service users who might have reported more conflicts than those who were re-interviewed. Despite these weaknesses, our analyses still offer new insights into conflict and user involvement and have potentially important implications for policy and practice.

As we discussed at the start of our paper, there has been an increased acceptance of user involvement within the addictions in recent years. However, involving drug users in making decisions about treatment has the potential to generate conflicts which could ultimately undermine the treatment process. Clients are often viewed as dishonest and manipulative, staff may sometimes actively provoke conflict with drug users as part of a broader therapeutic process, and some treatment agencies are hierarchically organised and rule-bound in ways that inhibit flexibility to individual needs. Beyond this, aspects of service provision are dictated by policies and funding mechanisms that transcend any particular service, and some treatment decisions (for example, dose levels in substitute prescribing) might be viewed as inherently contentious.

Perhaps unsurprisingly then, our study found evidence of conflict in both residential and community drug treatment settings, particularly during the initial detoxification stage of a residential programme when clients were withdrawing and feeling vulnerable, confused, stressed and isolated. The types of conflict identified were diverse, and ranged from relatively trivial differences of opinion through to more serious disagreements that resulted in clients leaving, or being told to leave, treatment prematurely. Although overt conflict involving shouting and verbal abuse was relatively infrequent, forms of 'latent' conflict–that is, negative experiences and problematic aspects of treatment that could potentially escalate into overt disputes–were common.

Notwithstanding the above, our study found little evidence that drug users were manipulative, overdemanding, aggressive or impatient. On the contrary, they seemed positive and co-operative at treatment entry and were often hoping and expecting to be guided by staff in treatment decision-making. Indeed, many wanted to be honest and open with the professionals who were helping them and appeared keen to discuss and resolve any problems that might arise. Beyond this, some clients believed that they had to accept staff decisions if they wanted to receive support and others adopted non-confrontational forms of covert resistance (rather than open discontent) when faced with treatment decisions about which they were unhappy.

Such findings are broadly consistent with sociological literature which has suggested that overt conflict between clients and healthcare professionals is rare [30-32] and clients will often behave in ways that demonstrate resistance to staff power, but without any actual open demonstration of dissent [33,34]. Furthermore, our data showed how staff were on occasions able to avoid conflict in treatment encounters by skilfully deploying user involvement as a strategy for soliciting client compliance. For example, staff sometimes offered reasoned grounds for treatment decisions in order to increase the likelihood of client adherence to treatment protocols; provided structured opportunities to discuss decisions on the governance of treatment facilities in order to increase client acceptance of the social structures of treatment; and provoked dispute through elaborate hierarchies and house rules in order to provide occasions for reality confrontation.

From this, we can conclude that involving drug users in decisions about their treatment–by, for example, listening to them, consulting with them, negotiating with them, questioning them, challenging them and even provoking them–does not automatically result in conflict. Moreover, when conflict (actual or latent) does occur, this will not necessarily be negative or detrimental to treatment processes or outcomes. However, we must also accept that conflict can be unproductive, may damage treatment relations and, in the worst case scenario, can lead to treatment breakdown. Accordingly, it is important to consider how user involvement might be implemented so that harmful forms of conflict can be minimised and avoided whenever possible.

In our research, negative conflict (latent and overt) often occurred because clients did not understand aspects of the treatment programme, particularly the reasoning behind rules and procedures. Equally, they became resentful when staff imposed different penalties on different clients for the same or similar rule violations. Such findings suggest that detrimental forms of conflict might often be avoided if staff spent more time communicating and explaining treatment issues to their clients. Supporting this, many drug users expressed a clear desire to talk to, and negotiate with, staff. However, in practice, they often failed to voice their concerns calmly and reasonably. Arguably, therefore, user involvement might also be improved if services could offer clients more basic training in communication skills, stress management and conflict resolution.

Drug users' criticisms of particular treatments (such as methadone), meanwhile, revealed that some individuals would rather abandon treatment altogether than accept a service to which they were categorically opposed. This implies that services are more likely to be successful if clients have some choice about, and involvement in deciding, what treatment/s they actually receive. Despite this, we live in a world of finite resources and there may be very good reasons why an individual cannot receive a particular form of support. Again, in such situations, negative conflict might be minimised or avoided if clients were given clearer explanations regarding why a particular treatment option could not be received and/or if staff focused more on trying to manage and moderate any unrealistic expectations which clients appeared to hold.

Interestingly, some clients acknowledged that their own bad behaviour and negative states of mind, including not really being 'ready' to address their drug problems, had caused conflicts with staff. Coping with argumentative drug users and ascertaining whether or not someone is genuinely psychologically and emotionally 'ready' for a particular treatment will inevitably be difficult and call for a high level of staff sensitivity, skill and understanding. Staff must also recognise that drug users are not inevitably disruptive and that those who are being difficult may simply be reacting to their own stressful life circumstances. This seems particularly important in residential services, especially when clients are undertaking detoxification.

Our findings have shown that staff do manage conflict situations, including sometimes successfully converting the client to the staff's point of view. Yet, it has also been evident that many conflict situations are not successfully managed, causing clients to feel angry and resentful, but also contributing to them breaching treatment protocols and leaving treatment prematurely. Since many drug users highlighted staff negativity towards them as a source of actual and latent conflict, those working in drug services might require more training to enable them better to understand, and be more sympathetic towards, the circumstances, motivations, and stresses that new treatment entrants commonly face. Arguably, if drug users felt more respected by service staff–particularly in the early stages of treatment–they might be more willing to engage positively with treatment protocols and feel and behave more like genuine partners in the treatment process.

Competing Interests

The authors declare that they have no competing interests.

Authors' Contributions

JF assisted in the late design stages of the study, collected and analysed data, and drafted the manuscript. MB and JN designed the study, analysed data, and participated in drafting and revising the manuscript. NJ collected and analysed data, and commented on the draft. All authors read and approved the final manuscript.

Acknowledgements

We thank the Joseph Rowntree Foundation for financial support; Lee Berney for taking part in research design and interviewing; Jenny Keen for advice and treatment agency liaison; Catherine J. Davison and the anonymous reviewers for their comments on the paper; and members of the Project Advisory Group (Elliot Albert, Robin Bunton, Joan Currie, Gemma Dunn, Nicola Edge, Jane Fountain, Charlie Lloyd [chair], Neil Hunt, Neil Hunter, Nick Manning and Jenny Scott) for suggestions and comments throughout study. The usual disclaimer applies. Last but not least, we thank the four drug treatment services for helping us to recruit to the study, as well as the many staff and clients who gave up their time to be interviewed.

References

1. Barker I, Peck E, (Eds): Power in Strange Places: User empowerment in Mental Health Services. London: Good Practices in Mental Health; 1987.

2. Davies A: Users' Perspectives. In Psychiatry in Transition: the British and Italian experiences. Edited by: Ramon S, Gianichedda MG. London: Pluto Press; 1988.
3. Taylor M, Hoyes L, Lart R, Means R: User Empowerment in Community Care: Unravelling the issues. Bristol: University of Bristol School for Advanced Urban Studies; 1992.
4. Thompson J: User Involvement in Mental Health Services: The Limits of Consumerism, the Risks of Marginalisation and the Need for a Critical Approach, Research Memorandum No. 8. Hull: The Centre for Systems Studies (University of Hull); 1995.
5. Small N, Rhodes P: Too Ill to Talk? User involvement and palliative care. London: Routledge; 2000.
6. Barnes M, Wistow G: Understanding user involvement. In Researching User Involvement. Edited by: Barnes M, Wistow G. Leeds: Nuffield Institute for Health Services Studies; 1992:1–15.
7. Croft S, Beresford P: From Paternalism to Participation: Involving People in Social Services. London: Open Services Project/Joseph Rowntree Foundation; 1990.
8. Crawford M, Rutter D, Manley C, Weaver T, Bhui K, Fulop N, Tyrer P: Systematic review of involving patients in the planning and development of health care. BMJ 2002, 325(7375):1263.
9. Fischer J, Jenkins N, Bloor M, Neale J, Berney L: Drug User Involvement in Treatment Decisions. York: Joseph Rowntree Foundation; 2007.
10. National Treatment Agency: NTA Guidance for Local partnerships on User and Carer Involvement. London: National Treatment Agency; 2006:11.
11. Graham K, Saunders S, Flower MC, Birchmore Timney C, White-Campbell M, Zeidman Pletropaolo A: Addictions Treatment for Older Adults: Evaluation of an Innovative Client-Centered Approach. New York: Haworth Press; 1995.
12. Kraybill K, Zerger S: Providing Treatment for Homeless People with Substance Use Disorders: Case Studies of Six Programs. Nashville, TN: National Health Care for the Homeless Council; 2003.
13. Friends of Addiction Recovery–New Jersey: Consumer Advisory Committee Progress Report: Towards a Client-Centered/Recovery-Oriented System of Care. Robbinsville, NJ. 2007.
14. Biase DV: Adolescent heroin abusers in a therapeutic community: Use of the MAACL emotional traits and splitting in treatment. Journal of Psychedelic Drugs 1971, 2:145–147.

15. De Leon G, Skodol A, Rosenthal MS: The Phoenix House therapeutic community for drug addicts: Changes in psychopathological signs. Archives of General Psychiatry 1973, 28:131–135.
16. Biase DV, De Leon G: The encounter group: Measurement of some affect changes. In Phoenix House: Studies in a therapeutic community (1968–1973). Edited by: De Leon G. New York: MSS Information Corporation; 1974:85–89.
17. Holland S: Psychiatric severity in the TC. In Bridging services: Drug abuse, human services and the therapeutic community–Proceedings of the 9th World Conference of Therapeutic Communities, September 1–6, 1985. Edited by: Acampora A, Nebelkopf E. San Francisco: Abacus Printing; 1986:122–131.
18. Nurco DN, Hanlon TE, O'Grady KE, Kinlock TW: The association of early risk factors to opiate addiction and psychological adjustment. Criminal Behaviour and Mental Health 1997, 7:213–228.
19. Pallone NJ, Hennessy JJ: Tinder-box criminal aggression. New Brunswick: Transaction Publishers; 1996.
20. De Leon G: The Therapeutic Community: Theory, Model, and Method. New York: Springer; 2000:159.
21. Neale J: Drug users' views of prescribed methadone. Drugs: Education, Policy & Prevention 1998, 5:33–45.
22. Garrett D, Foster J: Fumbling in the dark. Druglink 2005, 12.
23. Vaillant GE: Sociopathy as a human process: a viewpoint. Archives of General Psychiatry 1975, 32:178–183.
24. Wexler HK: The success of therapeutic communities for substance abusers in American prisons. J Psychoactive Drugs 1995, 27(1):57–66.
25. Polcin DL: Professional therapy versus specialised programs for alcohol and drug abuse treatment. Journal of Addiction and Offender Counselling 2000, 21:2–12.
26. Polcin DL: Rethinking Confrontation in Alcohol and Drug Treatment: Consideration of the Clinical Context. Substance Use & Misuse 2003, 38:165–184.
27. Morrice JK: Basic concepts: a critical review. In Therapeutic Communities: Reflections and Progress. Edited by: Manning N, Hinshelwood R. London: Routledge; 1979:49–58.
28. Yablonsky L: The Tunnel Back. New York: Macmillan; 1965.
29. Bloor M, McKeganey N, Fonkert D: One Foot in Eden: a Sociological Study of the Range of Therapeutic Community Practice. London: Routledge; 1988.

30. Bloor M, Horobin G: Conflict and conflict resolution in doctor/patient interactions. In A Sociology of Medical Practice. Edited by: Cox C, Meade A. London: Collier-Macmillan; 1975:271–284.
31. Stimson G, Webb B: Going to See the Doctor: the Communication Process in General Practice. London: Routledge; 1975.
32. Strong PM: The ceremonial order of the clinic: Parents, doctors and medical bureaucracies. London: Routledge & Kegan Paul; 1979.
33. Bloor M, McIntosh J: Surveillance and concealment: a comparison of techniques of client resistance in therapeutic communities and health visiting. In Readings in Medical Sociology. Edited by: Cunningham-Burley S, McKeganey N. London: Tavistock; 1989:159–181.
34. Goffman E: Asylums. New York:Doubleday Anchor; 1961.
35. Hughes R: Considering the vignette technique and its application to a study of drug injecting and HIV risk behaviour. Sociology of Health & Illness 1998, 20:381–400.
36. Ritchie J, Spencer L: Qualitative data analysis for applied policy research. In Analysing Qualitative Data. Edited by: Bryman A, Burgess RG. London: Routledge; 1994:173–194.
37. Cooley C: Human Nature and Social Order. London: Transaction; 1983. (original edition: 1902)

Deficits in Implicit Attention to Social Signals in Schizophrenia and High Risk Groups: Behavioural Evidence from a New Illusion

Mascha van't Wout, Sophie van Rijn, Tjeerd Jellema, René S. Kahn and André Aleman

ABSTRACT

Background

An increasing body of evidence suggests that the apparent social impairments observed in schizophrenia may arise from deficits in social cognitive processing capacities. The ability to process basic social cues, such as gaze direction and biological motion, effortlessly and implicitly is thought to be a prerequisite for establishing successful social interactions and for construing a sense of "social intuition." However, studies that address the ability to effortlessly process basic social cues in schizophrenia are lacking. Because social cognitive processing deficits may be part of the genetic vulnerability for schizophrenia, we also investigated two groups that have been shown to be at increased risk of developing schizophrenia-spectrum pathology: first-degree relatives of schizophrenia patients and men with Klinefelter syndrome (47,XXY).

Results

We compared 28 patients with schizophrenia, 29 siblings of patients with schizophrenia, and 29 individuals with Klinefelter syndrome with 46 matched healthy control subjects on a new paradigm. This paradigm measures one's susceptibility for a bias in distance estimation between two agents that is induced by the implicit processing of gaze direction and biological motion conveyed by these agents. Compared to control subjects, patients with schizophrenia, as well as siblings of patients and Klinefelter men, showed a lack of influence of social cues on their distance judgments.

Conclusions

We suggest that the insensitivity for social cues is a cognitive aspect of schizophrenia that may be seen as an endophenotype as it appears to be present both in relatives who are at increased genetic risk and in a genetic disorder at risk for schizophrenia-spectrum psychopathology. These social cue–processing deficits could contribute, in part, to the difficulties in higher order social cognitive tasks and, hence, to decreased social competence that has been observed in these groups.

Introduction

One of the cardinal dysfunctions associated with the schizophrenia phenotype concerns disturbances in social functioning [1]. Although some researchers have argued that this might be a consequence of severe psychopathology, others have demonstrated that social dysfunction is relatively independent of symptomatology [2]. This latter view is further supported by findings that disturbances in social functioning are already present in early adolescence and often precede the onset of psychosis [3]–[5]. In the search for determinants of social dysfunction in schizophrenia, adequate cognitive processing of social information appears to be of crucial importance. In the last decade a growing body of research demonstrated deficits in social information processing in schizophrenia [6], including difficulties in emotion recognition [7]–[9], an inability to understand and manipulate other people's behaviour in terms of their mental states, also called Theory of Mind, as well as an insensitivity to interpersonal social cues that refer to someone's affect and goals [10]. Interestingly, these social cue processing deficits seem to be independent of intelligence, i.e. not attributable to a generalized performance deficit [11], but are related to negative symptoms of schizophrenia, such as emotional withdrawal [12] and skills to perceive, process, and send social signs [13].

Indeed, the ability to quickly and effortlessly process social cues is an important underlying characteristic of successful social interactions and communication [14] as it allows a continuous interpretation of rapidly changing social signals. Examples of such basic social signs that are processed fast and effortlessly, or implicitly, are gaze direction, head orientation and body postures [15]. These cues can give clues about someone's intentions, goals and beliefs [16]. This fast and effortless processing of social cues may be especially relevant for construing a sense of 'social intuition' in which involuntary and implicit processes are crucial [17]. Intuitions have been described as follows: "intuitions are fast and take into account non-consciously generated information, gathered from experience, about the probabilistic structure of the cues and variables relevant to one's judgments, decisions, and behaviour" [18]. Although schizophrenia patients seem to fail in areas of social intuition and the implicit processing of social cues in social interactions as observed in their social behaviour [19], studies that address the ability to effortlessly process basic social cues in schizophrenia are scarce. The present study sought to remedy this, and examined the influence of the implicit processing of social cues on a distance judgment task in schizophrenia.

In addition to patients with schizophrenia and healthy controls, we included two other groups in the study: a) individuals at increased genetic risk for schizophrenia, i.e. siblings of schizophrenia patients and b) individuals with an X chromosomal disorder who are at risk for developing schizophrenia-like psychopathology, i.e. men with Klinefelter syndrome. Biological siblings of patients with schizophrenia have been shown to be at significantly higher risk for the development of schizophrenia [20], and display cognitive deficits similar to those observed in schizophrenia patients, although to a lesser degree [21]. Inclusion of the sibling group allowed us to study social cognitive deficits that are related to a genetic vulnerability to schizophrenia, without confounding environmental influences as hospitalization, medication and psychopathology. Support for the role of genetic mechanisms in social cognitive deficits comes from studies demonstrating abnormalities in the processing of social-emotional cues in biological relatives of patients with schizophrenia [22], [23]. This fits with the finding that social skills are under considerable genetic control in the general population.

The third experimental group consisted of men with Klinefelter syndrome who have an extra X-chromosome (47,XXY chromosomal pattern). Klinefelter syndrome has been associated with serious social difficulties and social cognitive deficits, such as high levels of social anxiety, communication difficulties and impaired social skills as well as deficits in interpreting non-verbal social signals [24]–[26].

Furthermore, high levels of schizotypal traits and schizophrenia symptoms have been observed in men with Klinefelter syndrome and include ideas of reference, unusual perceptual experiences, magical thinking, odd speech, disorganized thinking, suspiciousness and excessive social anxiety [27]–[30]. In addition, the life-time prevalence of psychotic disorders in XXY men appears to be 16 times higher as compared to men from the general population [30] and Klinefelter has been associated with an increased relative risk of being hospitalized with severe psychopathology such as schizophrenia spectrum pathology [28]. Moreover, the prevalence of the XXY chromosomal pattern is higher among men with schizophrenia. Taken together this has led others to propose that Klinefelter syndrome may serve as a genetic model for psychosis [29], [31]. Therefore, similar to relatives of schizophrenia patients, this genetic population can be considered a high risk population for the development of schizophrenia. Considering the social cognitive deficits, Klinefelter syndrome may specifically serve as a model for investigating the contribution of social perception impairments to schizophrenia symptoms.

An additional advantage arising from studying Klinefelter men is knowledge about the precise genetic aetiology of this syndrome, in contrast to the limited knowledge of the genetic underpinnings of social cognitive impairments in schizophrenia. It has been hypothesized that the X chromosome may harbour genes that are crucially involved in development of the social brain [32]. Similarities between patients with schizophrenia, their siblings and XXY men might point to a role of the X-chromosome in the development of cognitive systems that are important for processing basic social signals [33] and because social perception deficits in Klinefelter syndrome may result from an X chromosomal abnormality, this may have heuristic value in the search for the genetic mechanisms underlying social perception deficits in schizophrenia. Indeed, there is reason to suspect the involvement of sex chromosomes as it might explain, at least in part, the sex differences that have been observed in social cognitive skills in the general population as well as in schizophrenia populations [34].

The aim of the present study was to investigate the implicit processing of basic social cues in three groups on the schizophrenia continuum, i.e. in individuals with schizophrenia, individuals with an increased genetic risk for schizophrenia and individuals with a genetic disorder who show schizophrenia-like symptoms. To this end we used a new paradigm involving a bias in the judgment of the distance between two agents induced by the implicit, i.e. effortless processing of social cues conveyed by these agents. In this task the social cues consisted of the direction of attention (gaze direction) and implied biological motion (body postures). We choose these social cues based upon an extensive body of research showing that biological motion can be accurately and effortlessly perceived [35]–[39] and that direct gaze serves as a precursor to social interaction [40]–[42].

Hence, these social cues can induce the sensation of people (dis)-engaging in a social interaction when their gaze or body postures attend towards (or away from) each other. Consequently, the automatic or implicit processing of gaze direction and implied biological motion can result in people judging the agents as closer together, compared to reference objects, whilst objectively this is not the case (see Jellema et al., 2004 for published pilot data in form of an abstract).

We hypothesized that patients with schizophrenia would demonstrate difficulties in the effortless or implicit processing of social cues compared to control participants, i.e. patients may show no response bias congruent with the direction of the social cues whereas this would be the case in control participants. We expected a similar lack of response bias in siblings of patients with schizophrenia and XXY men albeit to a lesser extent compared to patients with schizophrenia. Furthermore, we investigated the relationship between schizophrenia symptomatology and social cue processing in patients with schizophrenia. We predicted that lack of social cue processing would be especially prevalent in patients with negative symptoms, since these patients are in particular characterized by social-emotional disturbances.

Materials and Methods

Ethics Statement

The local ethics committee, METC-UMCU approved the study and all subjects provided written informed consent after the procedure had been fully explained according to Declaration of Helsinki.

Participants

33 Patients (23 men, 10 women) with a diagnosis of schizophrenia were recruited at the University Medical Centre Utrecht. All patients met the DSM-IV criteria for schizophrenia, as confirmed by the Comprehensive Assessment of Symptoms and History interview (CASH) [43] semi-structured interview designed for research in the major psychoses and was administered by a psychiatrist. Patients were also screened for affective disorders, i.e. depression and mania, and substance-related disorders, with the CASH. Most patients were diagnosed with paranoid schizophrenia (n = 22), one with disorganized type, one with residual type, six with undifferentiated type and three with schizophreniform disorder. Patients were clinically stable; four patients were inpatients and in remission and 29 were outpatients. 31 Patients received medication (30 patients only antipsychotics, such as leponex (n = 13), quetiapine (n = 4), olanzapine (n = 6), risperidone (n = 8) and one patient also received oxazepam). Symptoms and severity were independently rated by two raters with the Positive and Negative Syndrome Scale (PANSS) [44].

Raters were trained by a qualified trainer and followed inter-rater reliability training every six months. Mean positive symptoms was 14.22 (SD 5.22, range 7–27), negative symptoms 14.84 (SD 5.78, range 7–29) and general psychopathology 26.66 (SD 6.84, range 17–47). Mean duration of illness was 9.44 years (SD 8.01) and mean age of onset was 23.83 years (SD 5.45).

32 Siblings of patients with schizophrenia (12 men, 20 women) were recruited through advertisements at the Ypsilon website, which is a website dedicated to relatives of patients with schizophrenia. The diagnosis of schizophrenia for the affected sibling was confirmed with a CASH interview. However, due to ethical reasons we were unable to verify the diagnosis of schizophrenia for 12 affected siblings with the CASH interview.

32 Men with Klinefelter syndrome (47,XXY) were studied. The participants were recruited from the Dutch Klinefelter Association, and were not selected for psychological, behavioural or cognitive abnormalities. Diagnosis of Klinefelter syndrome was confirmed by karyotyping, using standard procedures. 50 Non-psychiatric control participants (31 men, 19 women) were drawn from the general population via advertisements in local newspapers.

Inclusion criteria for all participants were age between 18 and 65 years and good physical health. Exclusion criteria were neurological conditions, history of head injury with loss of consciousness, recent history of alcohol and substance abuse, or mental retardation. None of the control participants and siblings had a history of psychiatric illness or use of psychiatric medication confirmed with the Mini International Neuropsychiatric Interview plus [45]. The Dutch translation of the National Adult Reading Test (NART) [46] and Raven's Advanced Progressive Matrices [47] were used to match the groups on estimates of verbal and performance intelligence level, respectively [48]. See Table 1 for demographic data of participants that were included in the analyses as some participants were excluded from the analyses due to attentional problems (see also methods social distance judgment task and statistical analyses).

Table 1. Demographic data (mean (SD)) of participants included in the Social Distance Judgment Task analysis.

Variable	Patients	Siblings	Klinefelter men	Controls	P^1	P^2
N	28	29	29	46		
Age in years	32.4 (7.5)	34.6 (10.7)	38.1 (8.5)	31.9 (9.2)	0.45	0.08
Male:female ratio	18:10	11:18	1:0	27:20	0.11	NA
Education in years	14.3 (2.8)	16.2 (1.9)	13.9 (2.7)	14.9 (2.6)	0.01	0.56
Parental education in years	13.9 (2.9)	14.6 (2.7)	NA	13.2 (2.9)	0.27	NA
NART	103.6 (8.2)	104.5 (8.1)	102.7 (8.6)	107.6 (9.5)	0.13	0.09
Raven's Matrices	NA	109.2 (9.9)	107.7 (14.4)	108.4 (13.8)	0.79	0.24

P^1: Between-group comparisons of patients with schizophrenia, siblings of patients and control participants with ANOVA, except male:female ratio is analyzed with non-parametric Kruskal Wallis test, df = 100; P^2: Between-group comparisons of Klinefelter men and male controls with Student's t-test, df = 52; NA = Not available.

Social Distance Judgment Task

The Social Distance Judgment Task measures the illusion of de- or increasing distance caused by the implicit, or effortless, processing of social cues. The underlying principle behind the task is that the perceived distance between two agents will be influenced by social cues conveyed by these agents in comparison to the perceived distance between two geometrical objects that do not signal social intentions, even though the actual distance between the two agents and the two geometrical objects is the same. The social cues signalled by the two agents will result in a response bias paralleling the strength of social cues, i.e. the more social cues are present the stronger the bias will be. This had been confirmed previously in a pilot study (published abstract Jellema et al., 2004).

Stimuli were pairs of cartoon figures shown in running postures conveying two different social cues: gaze direction (figures looking away or towards each other) and biological motion (figures running away or towards each other). Head and body of the cartoon figures were pointing in the same direction, or in opposite directions, amounting to a total of four different compositions of cartoon figures, see Figure 1, top panel. One (male) cartoon figure was used, selected from the CorelDraw graphical package. Cartoon figures were always presented in pairs as each other's mirror-image (as displayed in Figure 1, bottom panel). All faces had the same, fairly neutral, expression.

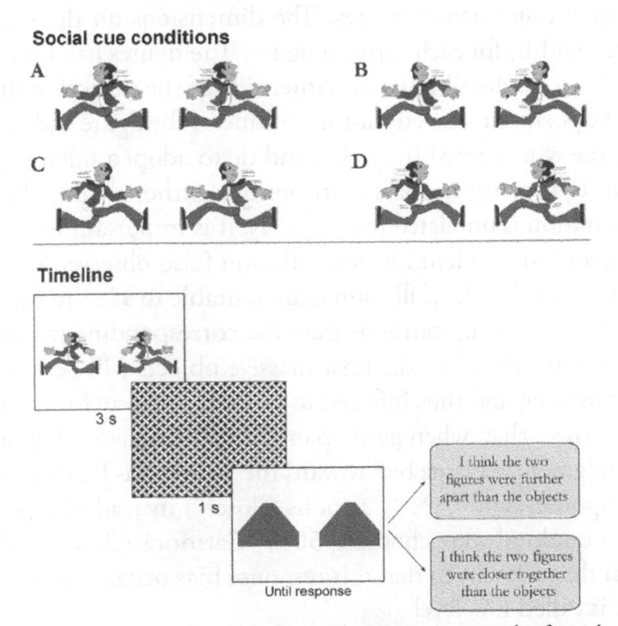

Figure 1. Response to social cues. Top panel: Left to right: increasing strength of social cues leading to the response: 'Cartoons Closer.' Bottom panel: Example of a single trial.

A pair of cartoon figures was presented for 3 s, after which a mask of 1 s was shown, followed by a pair of geometrical figures (see Figure 1, bottom panel, for an example of a trial). Participants had to choose one of two possible responses: (1) 'I think the two cartoon figures were closer together than the two geometrical objects' and (2) 'I think the two cartoon figures were further away from each other than the two geometrical objects.' For convenience, we labelled response 1 as 'Cartoons Closer' and response 2 as 'Cartoons Farther.' We chose to use this forced-choice paradigm to increase the likelihood to detect a response bias.

The task consisted out of 30 trials evenly distributed over the four social cue levels and so-called catch-trials, resulting in every social cue level occurring six times. Except for the six catch-trials, the distance between the geometrical figures was always the same as the distance between the cartoon figures. These catch-trials were used to allow exclusion of those participants from analysis that did not pay proper attention to the task. Participants who made more than two errors in the catch-trials were excluded from the analyses. Three different distances of 2, 3 or 4 cm between cartoons and geometrical figures were randomly presented. In the catch-trials there was a 2 cm difference between the geometrical figures and the cartoon figures. Before the onset of the task, participants completed six practice trials.

The maximal height and width of the geometrical objects matched those of the corresponding cartoon figures. The dimensions on the screen were 4.8×6.5 cm (height×width) for each cartoon figure. The figures had been digitally adapted such that the mass distribution on either side of the vertical midline was identical, with the eye positioned exactly at the midline of the figure and centred in the head. However, the wind caused the jacket and tie to adopt a sideway position resulting in a slight asymmetry in mass distribution. Although this slight asymmetry in mass distribution is unrelated to this study, it is important to briefly mention as it resulted in an independent low-level illusion (also observed in pilot data: Jellema et al., 2004). This low-level illusion is attributable to a less massive appearance due to spaces in the running cartoons than the corresponding pyramidal-shaped geometrical objects. Therefore the least massive objects will be judged furthest away from the observer, and thus inferred to be furthest away from each other. For that reason we expect that, when participants are not influenced by social cues in their distance judgments, a large bias toward the "Cartoons Farther" response (roughly 75% of responses, and 25% "Cartoons Closer") instead of expecting participants to respond randomly, i.e. choosing 50% "Cartoons Closer" and 50% "Cartoons Father" on the task. Given that this response bias occurs irrespective of social cues this effect is called low-level.

Statistical Analysis

Data from the Social Distance Judgment Task was analyzed with General Linear Model repeated measures ANOVA of within subject contrast with increasing social cue strength as a within subjects variable (four strength levels A to D, see Figure 1, top panel). The order of social cue conditions was based on pilot data demonstrating that the biological motion cue had a stronger effect on distance judgment than the gaze direction cue (published abstract Jellema et al., 2004). Our pilot data further revealed that the gaze direction cue facilitated the motion cue (or the other way around) to have an effect, i.e. both social cues are required to get a stronger visual illusion. Given the incongruency of biological motion and gaze direction in condition B and C (see Figure 1, top panel), we thus did not expect a significant difference in distance judgments between these conditions. Repeating the analyses with only three levels of social cues strength (A, BC, D) did not alter the results described below.

The repeated measures ANOVA was first done separately for the different groups (controls, patients, siblings, Klinefelter men) to investigate whether there was a significant linear increase consistent with social cue strength in each of the groups. Second a similar GLM repeated measures of within subject contrast with the four social cue strength levels as a within subjects variables, but with Group (control vs. experimental groups: patients, siblings) as between subject factor tested for differences between the groups on the influence of social cues on distance judgment. Because only males are affected with Klinefelter syndrome, we performed a separate GLM repeated measures analysis in which the between subjects factor Group consisted out of Klinefelter men versus male controls only.

Five patients with schizophrenia, three siblings, three Klinefelter men and four control participants made more than two errors in the catch-trials and were not included in further analyses. See Table 1 for demographic data of participants included in the analyses.

Results

Across all groups we observed the presence of a low-level effect that is noticeable in the general large bias toward the "Cartoons Farther" response compared to the "Cartoons Closer" response across all conditions and irrespective of social cues.

Social Distance Judgment Task: Control Subjects, Patients, and Siblings

First to determine whether we found an effect of social cues on distance judgments we examined performance in the control group using a repeated measure

ANOVA with the four social cues as within-subjects factors. This analysis revealed a significant main effect of the different levels of social cues (F(3,43) = 7.42, p = 0.006, indicating that the different social cues had different effects on the distance judgments in the task. A post hoc t-test confirmed that the two extreme conditions, i.e. condition A vs. D (see Figure 1, top panel for a reference to the different conditions), differed significantly from each other in distance estimations, t(45) = −3.63, p = 0.001. Therefore, we tested whether there was a significant linear increase in the percentage of response 'Cartoons Closer' with increasing social cue strength and this was indeed what we found, F(1,45) = 14.27, p = 0.0005. This shows that there was an influence of social cue strength on distance judgments according to the social cues.

We repeated the same analyses for the patient group, but did not observe a significant main effect of a linear increase in the percentage of the response 'Cartoons Closer' with increasing social cue strength, F(1,27) = 0.34, p = 0.56. This shows that the percentage response 'Cartoons Closer' did not change with increasing social cue strength in the patient group. Post hoc t-test confirmed a lack of significant difference between the congruent social cue conditions A vs. D in the patient group, t(27) = 0.43, p = 0.67, and in the incongruent social cue conditions B vs. C, t(27) = 0.39, p = 0.699.

Remarkably, this absence of a significant main effect of a linear increase according to social cue strengths, and thus the suggested absence of a response bias, was also found in the sibling group, F(1,28) = 0.77, p = 0.39. Again post hoc t-tests did not show a significant difference between social cue conditions A vs. D, t(28) = −0.77, p = 0.45, or between social cue conditions B vs. C, t(28) = −0.59, p = 0.56.

To test whether the pattern on the task was different for the three groups we performed a repeated measure ANOVA with the four social cues as within-subjects factors and group (patients, siblings and controls) as between-subject factor. There was a significant main effect of a linear response due to increasing social cue strength, F(1,100) = 4.33, p = 0.04. In addition, we observed a significant interaction between the groups and levels of social cue strength, F(2,100) = 3.79, p = 0.026, demonstrating that the pattern of the response 'Cartoons Closer' in proportion to social cue strengths differed between patients with schizophrenia, siblings and control subjects. See Figure 2.

Indeed, post-hoc tests revealed that the control group differed significantly from the patient group in sensitivity for social cues (F(1,72) = 8.06, p = 0.006), which was specifically due to a significant difference between social cue condition A vs. D, t(73) = −2.66, p = 0.01. The sibling group did not differ from the control group (F(1,73) = 2.21, p = 0.14), nor from the patient group (F(1,55) = 1.09, p = 0.30).

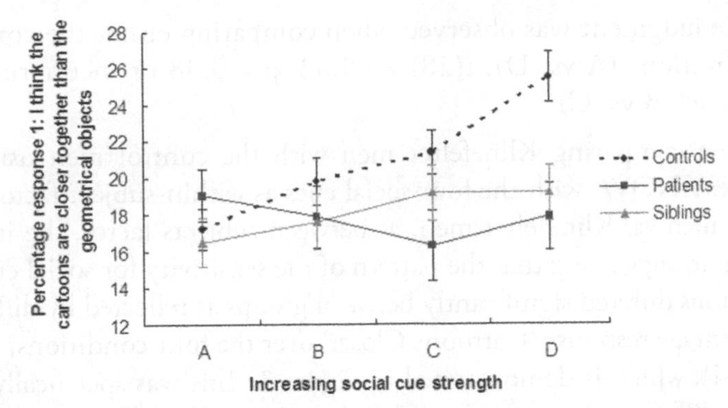

Figure 2. Linear increase in response 1 ("Cartoons Closer") consistent with social cue strength in healthy control subjects, but not in patients or sibling of patients.

Previous studies demonstrated sex differences in social-emotional information processing in schizophrenia. Although there was no significant difference between males and females in the control ($F(1,44) = 0.61$, $p = 0.44$) and sibling group ($F(1,27) = 1.27$, $p = 0.27$), we observed a trend in the schizophrenia patients ($F(1,26) = 3.23$, $p = 0.08$). In which in particularly male patients showed an abnormal pattern (although the influence on social cues was also not significant for female patients).

Social Distance Judgment Task: Klinefelter Men and Control Men

Again we first wanted to confirm the presence of an effect of social cue condition on distance judgements in the male control group. We performed a repeated measures ANOVA with four social cues as within-subjects factors as previously performed for the control, patient and sibling group separately. We found a significant main effect of a response bias congruent with the strength of the social cues, i.e. a significant linear increase in underestimations (i.e. increase in percentage response 'Cartoons Closer') of the perceived distance as strength of the social cues increased, $F(1,24) = 13.54$, $p = 0.001$. As before there was a significant difference on distance judgment between the two congruent social cue conditions A vs. D, $t(24) = -3.70$, $p = 0.001$, whereas distance judgments in the incongruent conditions, B vs. C, did not differ significantly from one another, $t(24) = -0.15$, $p = 0.88$.

In Klinefelter men on the other hand, the main effect of the repeated measures ANOVA was non-significant, suggesting that the percentage of response 'Cartoons Closer' did not change with increasing strength of the social cues, $F(1,28) = 0.001$, $p = 0.98$. Consequently, no significant difference in distance judgment was observed when comparing either the congruent social cue conditions (A vs. D), $t(28) = -0.02$, $p = 0.98$ or incongruent social cue conditions (B vs. C).

When comparing Klinefelter men with the control men using a repeated measures ANOVA with the four social cues as within-subject factors and Group: control men vs. Klinefelter men, as between-subjects factor, the interaction was significant, suggesting that the pattern of the sensitivity for social cues in distance estimations differed significantly between groups as reflected by different patterns of percentage response 'Cartoons Closer' over the four conditions, $F(1,52) = 4.4$, $p = 0.04$), which is demonstrated in Figure 3. This was specifically due to a significant difference between condition A vs. D, $t(52) = -2.10$, $p = 0.04$. Although potential age differences could have influenced the results, we did not find an effect of age on task performance, and the difference in pattern for sensitivity for social cues between the Klinefelter group and control group remained significant ($p = 0.04$).

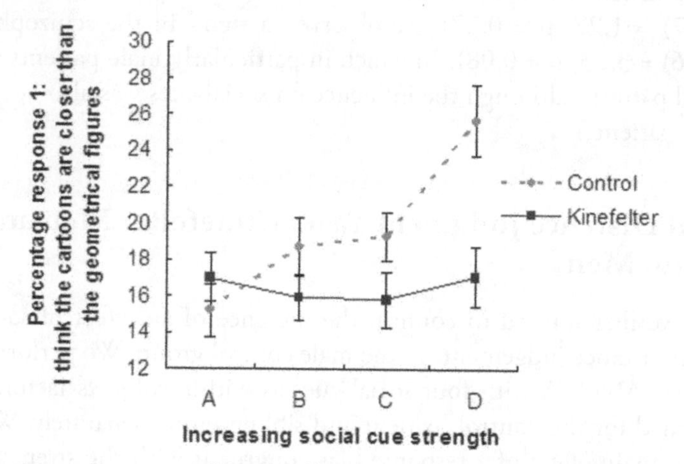

Figure 3. Linear increase in response 1 ("Cartoons Closer") consistent with social cue strength in healthy control men, but not in Klinefelter men.

Social Distance Judgment and Symptomatology

There was a significant negative correlation between the response bias due to social cue strength and negative symptoms of schizophrenia as measured with the PANSS, Spearman's rho = -0.47, $p = 0.01$. This suggests that patients with more negative symptoms are less influenced by social cues. There were no significant correlations between positive symptoms or general psychopathology as measured with the PANSS and influence of social cues.

Discussion

This study examined the implicit, or effortless processing of basic social cues, i.e. biological motion and gaze direction in three different groups: a) schizophrenia patients, b) individuals at increased genetic risk for schizophrenia, i.e. siblings of schizophrenia patients and c) individuals with an X-chromosomal disorder and high levels of schizotypal traits, i.e. men with Klinefelter syndrome.

In healthy controls, an increasing strength of social cues in the stimuli was accompanied by an increasing illusion of the perceived distance between the stimuli, indicating that social cues affected distance judgments as we expected. In contrast, in schizophrenia patients, siblings of patients and Klinefelter men, an increasing strength of social cues in the stimuli did not have any effect on the perceived distance between the stimuli, indicating that these social cues were not incorporated in the process of judging the stimuli. As a consequence these participants were less biased by social cues in their judgments and thus appear to be more accurate in their distance judgments as compared to controls. However, all groups did show a low level illusion caused by differences in mass distribution of the objects, suggesting a specific insensitivity to social cues in the experimental groups.

When considering the groups separately, schizophrenia patients and Klinefelter men were less sensitive to social cues, as compared to controls. Performance of the siblings of patients was in between patients and control participants; that is, siblings did not differ significantly from either controls or patients. The differences are probably not due to differences in general cognitive functioning, as the groups were matched on (parental) education and measures of intelligence. Furthermore, the subjects included in the analysis understood the task and were able to perform the task correctly, as indicated by their small number or absence of errors on the catch trials. In addition, the inclusion of subjects with no or very low numbers of errors on the catch trials rule out the possibility of visuospatial disabilities as well as attentional deficits that would make completion of the task difficult.

The current results suggested that patients with schizophrenia demonstrate a lack of sensitivity to even basic, simple social cues, in addition to deficits in more abstract, higher-order social cue recognition, as suggested by Corrigan and Green [10]. A failure to involuntary (implicitly) and quickly process these basic social cues may contribute to difficulties in social intuition, and hence in coping with social situations in these patients. Also, because less basic social information is available, more widespread effects on ('upstream'-) higher-order social cognitive processing can be expected. The observed insensitivity to social cues may underlie social cognitive deficits and social dysfunction in schizophrenia. The implicit processing of social cues is thought to be especially important for the forming of a Theory of Mind, i.e. the ability to infer someone's intentions, goals and beliefs

[16] and deficits in the effortless processing of these social cues might lead to disturbances in the attribution of mental states to others [14]. Indeed, a recent study demonstrated that patients with schizophrenia were impaired in using appropriate language to describe Theory of Mind animations [49].

Our results showed that especially patients with negative symptoms, which comprise social and emotional withdrawal, were insensitive to the influence of the social cues in their judgments. Patients with negative symptoms typically show problematic social functioning [50]–[52], but also deficits in other social emotional tasks [53]–[56]. Thus, these results corroborate previous research demonstrating that patients with schizophrenia show deficits in the processing of social information [6], with more severe impairments in patients with negative symptoms [56]–[59], though future research is needed to examine if any particular symptoms of the negative subscale is related to difficulties in the effortless processing of social cues. However, this study extends previous research in demonstrating deficits in the normally effortless processing of simple social cues.

Interestingly, the absence of an influence of the social cues on distance judgments was also observed in individuals at increased genetic risk for schizophrenia (siblings of patients) and in individuals with a genetic disorder associated with increased schizophrenia spectrum pathology (Klinefelter syndrome). Based on these findings three important conclusions can be drawn. First, siblings as well as the Klinefelter men were not clinically psychotic and did not use antipsychotic medication. The lack of sensitivity for social cues could thus not be due to the effects of illness or the medication use. In that way, these results validate the observed results in patients. Second, we propose that the observed lack of sensitivity for social cues is related to a genetic vulnerability to schizophrenia. The results showed that there were no differences between patients and siblings in distance judgment, suggesting that siblings resemble patients in their absence of implicit processing of social cues. However, it is important to note that the sibling group also did not differ from the control group and one could as well argue that siblings performed comparable to the control group.

Nevertheless, when taking the within group analysis into account we demonstrated that siblings, in contrast to controls, did not show a linear increase in underestimations, i.e. their distance judgments were not influenced by the social cues of human figures running towards each other or looking towards each other. Thus, our findings imply that the performance of siblings resembles the lack of sensitivity to social cues observed in schizophrenia patients, albeit to a lesser extent. Moreover, our results mirror and extend previous studies demonstrating impairments in other types of social emotional cue processing in relatives of patients with schizophrenia such as recognizing emotional facial expressions [22], [23], suggesting that problems in social cue processing might be regarded as a genetic vulnerability for schizophrenia. Third, additional evidence for a genetic loading on social cue processing comes from the finding in individuals with Klinefelter who show a similar lack of social cue processing on distance judgments as patients. As this disorder is defined by an X chromosomal abnormality, impaired cognitive processing of social cues in this group can be regarded as the expression of X-linked genetic pathology. Klinefelter men also display impairments in higher order social cognitive processing, such as recognition of facial expressions and emotional prosody, i.e. tone of voice [25]. The present findings suggest that the insensitivity to social cues could be regarded as an endophenotype that is shared by schizophrenia patients and Klinefelter men. Hence, not only in Klinefelter syndrome, but also in the schizophrenia spectrum, we might consider a role of X-linked genetic pathology underlying impairments in effortless processing of social information. This might explain, at least in part, the sex differences that have been observed in the incidence and severity in schizophrenia [34], [60], although we only observed a trend for male schizophrenia patients to be less sensitive to social cues compared to female schizophrenia patients. Moreover, this cognitive endophenotype may may also be present in other psychiatric disorders characterized by social cognitive deficits and the recognition of endophenotypes can contribute to the early detection of and possibly preventive treatment for certain psychiatric disorders.

With regard to the neural correlates involved in the processing of biological motion and social attention, the superior temporal gyrus, medial prefrontal cortex and anterior cingulate have been implicated [15]. Both in schizophrenia patients as well as relatives, abnormalities in these regions have been reported [61]–[67]. Interestingly, structural abnormalities in the anterior cingulate and the superior temporal gyrus have been found in Klinefelter syndrome as well [68]. Future studies should relate neural substrates of social cue processing in schizophrenia and relatives together with measures of social functioning. This would elucidate the relationship between the ability to process social cues and social behaviour and its underlying brain pathology in schizophrenia and provide more insight into the biological vulnerability to schizophrenia.

Finally, it is important to note some limitations of this study. For instance, it would be interesting to include a patient control group that is not associated with an increased risk to develop schizophrenia to demonstrate that this patient group is indeed susceptible to the illusion. In addition, a non-social condition could be included in the task to substantiate that the absence of the illusion is specific for social cues. Especially because previous research on visual illusions in schizophrenia has shown a reduced susceptibility in schizophrenia [69]–[71] and thus the current results need to be interpreted with caution. Nevertheless, we did observe the presence of a low-level illusion in all experimental groups demonstrating that patients with schizophrenia, their siblings and XXY men are susceptible to some perceptual illusions unrelated to these social cues. Further this highlights that other cognitive deficits, such as working memory, attention or visual deficits, probably do not explain our results. Another issue concerns a possible selection bias in the Klinefelter group. Since many men with Klinefelter syndrome remain undiagnosed [72] and untreated, the present results might not generalize to the general Klinefelter population. Finally our results might have been different if we had used more realistic stimuli in which decoding of social cues is more relevant instead of cartoon figures.

In summary, this study investigated the influence of simple, usually implicitly processed, basic social cues, i.e. biological motion and gaze direction, on distance judgements in individuals with a) a diagnosis of schizophrenia b) an increased risk for schizophrenia (siblings of patients) and c) with a genetic disorder associated with increased schizophrenia spectrum pathology (Klinefelter syndrome). Results showed that patients with schizophrenia, siblings of patients with schizophrenia and Klinefelter men (47, XXY) did not process these social cues effortlessly (involuntary or implicitly) compared to healthy controls. Within the schizophrenia group, this was especially the case in patients with more severe negative symptoms, i.e. patients that show additional social emotional disturbances. Hence, social cue processing deficits seem related to the vulnerability for schizophrenia, instead of illness in general and with a potential involvement of genes on the X chromosome. These basic social cue processing deficits might underlie impairments in other aspects of social cognition and social functioning. Future research should investigate the relationships among insensitivity to social cues, social functioning and neurobiological substrates in schizophrenia as well as schizotypal symptoms in high-risk groups.

Acknowledgements

We would like to thank E. Caspers and W. Cahn for help with the recruitment of patients and T. Rietkerk for help with data acquisition.

Authors' Contributions

Conceived and designed the experiments: MvW SvR TJ RSK AA. Performed the experiments: MvW SvR. Analyzed the data: MvW SvR. Contributed reagents/materials/analysis tools: MvW TJ AA. Wrote the paper: MvW SvR TJ RSK AA.

References

1. A.P.A. (1994) Diagnostic and Statistical Manual of Mental Disorders (4th ed.). Washington, DC: American Psychiatric Association Press.
2. Lenzenweger MF, Dworkin RH (1996) The dimensions of schizophrenia phenomenology. Not one or two, at least three, perhaps four. Br J Psych 168: 432.
3. Hans SL, Marcus J, Henson L, Auerbach JG, Mirsky AF (1992) Interpersonal behavior of children at risk for schizophrenia. Psychiatry 55: 314–335.
4. Walker EF (1994) Developmentally moderated expressions of the neuropathology underlying schizophrenia. Schizophr Bull 20: 453–480.
5. Baum KM, Walker EF (1995) Childhood behavioral precursors of adult symptom dimensions in schizophrenia. Schizophr Res 16: 111–120.
6. Pinkham AE, Penn DL, Perkins DO, Lieberman J (2003) Implications for the Neural Basis of Social Cognition for the Study of Schizophrenia. Am J Psychiatry 160: 815–824.
7. Edwards J, Jackson HJ, Pattison PE (2002) Emotion recognition via facial expression and affective prosody in schizophrenia: a methodological review. Clin Psychol Rev 22: 789–832.
8. Kohler CG, Brennan AR (2004) Recognition of facial emotions in schizophrenia. Curr Opin Psychiatry 17: 81.
9. Van't Wout M, Aleman A, Kessels RPC, Cahn W, de Haan EHF, et al. (2007) Exploring the nature of facial affect processing deficits in schizophrenia. Psychiatry Res 150: 227–235.
10. Corrigan PW, Green MF (1993) Schizophrenic patients' sensitivity to social cues: The role of abstraction. Am J Psychiatry 150: 589–594.
11. Corrigan PW (1994) Social cue perception and intelligence in schizophrenia. Schizophr Res 13: 73–79.
12. Corrigan PW, Green MF, Toomey R (1994) Cognitive correlates to social cue perception in schizophrenia. Psychiatry Res 53: 141–151.
13. Corrigan PW, Toomey R (1995) Interpersonal problem solving and information processing in schizophrenia. Schizophr Bull 21: 395–403.
14. Frith CD, Frith U (1999) Interacting Minds–A Biological Basis. Science 286: 1692–1695.
15. Jellema T, Perret DI (2005) Neural Basis for the Perception of Goal-Directed Actions. In: Easton A, Emery NJ, editors. The Cognitive Neuroscience of Social Behavior. New York: Psychology Press.

16. Perrett DI (1999) A cellular basis for reading minds from faces and actions. In: Hauser M, Konishi M, editors. The design of animal communication. Cambridge, MA: The MIT Press.
17. Lieberman MD (2000) Intuition: A social cognitive neuroscience approach. Psychol Bull 126: 109–136.
18. Bruner J (1960) The process of education. Cambridge, MA: Harvard University Press.
19. Bellack AS, Morrison RL, Wixted JT, Mueser KT (1990) An analysis of social competence in schizophrenia. The British Journal Of Psychiatry: The Journal Of Mental Science 156: 809–818.
20. Gottesman II (1991) Schizophrenia genesis: the origin of madness. New York: Freeman.
21. Sitskoorn M, Aleman A, Ebisch S, Appels M, Kahn RS (2004) Cognitive deficits in relatives of patients with schizophrenia: a meta-analysis. Schizophr Res 71: 285–295.
22. Toomey R, Seidman LJ, Lyons MJ, Faraone SV, Tsuang MT (1999) Poor perception of nonverbal social-emotional cues in relatives of schizophrenic patients. Schizophr Res 40: 121–130.
23. Loughland CM, Williams LM, Harris AW (2004) Visual scanpath dysfunction in first-degree relatives of schizophrenia probands: evidence for a vulnerability marker? Schizophr Res 67: 11–21.
24. Van Rijn S, Aleman A, Swaab H, Krijn T, Vingerhoets G, et al. (2007) What is said versus how it is said: Comprehension of affective prosody in men with Klinefelter (47,XXY) syndrome. J Int Neuropsychol Soc (JINS) 1065–1070.
25. Van Rijn S, Swaab H, Aleman A, Kahn RS (2006) X Chromosomal effects on social cognitive processing and emotion regulation: A study with Klinefelter men (47,XXY). Schizophr Res 84: 194–203.
26. Van Rijn S, Swaab H, Aleman A, Kahn RS (2008) Social behavior and autism traits in a sex chromosomal disorder: Klinefelter (47XXY) syndrome. J Autism Dev Disord 38: 1634–1641.
27. Van Rijn S, Aleman A, Swaab H, Kahn R (2006) Klinefelter's syndrome (karyotype 47,XXY) and schizophrenia-spectrum pathology. Br J Psychiatry 189: 459–461.
28. Bojesen A, Juul S, Birkebaek NH, Gravholt CH (2006) Morbidity in Klinefelter syndrome; a Danish register study based on hospital discharge diagnoses. J Clin Endocrinol Metab 91: 1254–1260.

29. DeLisi LE, Maurizio AM, Svetina C, Ardekani B, Szulc K, et al. (2005) Klinefelter's syndrome (XXY) as a genetic model for psychotic disorders. Am J Med Genet B Neuropsychiatr Genet 135: 15–23.
30. Boks MPM, de Vette MHT, Sommer IE, van Rijn S, Giltay JC, et al. (2007) Psychiatric morbidity and X-chromosomal origin in a Klinefelter sample. Schizophr Res 93: 399.
31. DeLisi LE, Friedrich U, Wahlstrom J, Boccio-Smith A, Forsman A, et al. (1994) Schizophrenia and sex chromosome anomalies. Schizophr Bull 20: 495–505.
32. Skuse D (2003) X-linked genes and the neural basis of social cognition. Autism: Neural Basis and Treatment Possibilities, Novartis Foundation Symposium 251: 84–98.
33. Skuse D, Morris JS, Dolan RJ (2005) Functional dissociation of amygdala-modulated arousal and cognitive appraisal, in Turner syndrome. Brain 128: 2084–2096.
34. Scholten MRM, Aleman A, Montagne B, Kahn RS (2005) Schizophrenia and processing of facial emotions: Sex matters. Schizophr Res 78: 61–67.
35. Johansson G (1973) Visual perception of biological motion and a model for its analysis. Perception & Psychophysics 14: 201–211.
36. Cutting JE, Kozlowski LT (1977) Recognising friends by their walk: gait perception without familiarity cues. Bull Psychon Soc 9: 353–356.
37. Sumi S (1984) Upside-down presentation of the Johansson moving light-spot pattern. Perception 13:
38. Dittrich WH (1993) Action categories and the perception of biological motion. Perception 22.
39. Kozlowski LT, Cutting JE (1977) Recognizing the sex of a walker from point-lights display. Perception & Psychophysics 21.
40. Argyle M, Cook M (1976) Gaze and mutual gaze. Cambridge: Cambridge University Press.
41. Kleinke CL (1986) Gaze and eye contact: a research review. Psychol Bull 100: 78–100.
42. Baron-Cohen S (1995) Mindblindness: An essay on autism and Theory of Mind. Cambridge: MIT Press.
43. Andreasen NC, Flaum M, Arndt S (1992) The Comprehensive Assessment of Symptoms and History (CASH) - an Instrument for Assessing Diagnosis and Psychopathology. Arch Gen Psychiatry 49: 615–623.

44. Kay SR, Fiszbein A, Opler LA (1987) The Positive and Negative Syndrome rating Scale (PANSS) for Schizophrenia. Schizophr Bull 13: 261–276.
45. Sheehan DV, Lecrubier Y, Sheehan KH, Amorim P, Janavs J, et al. (1998) The Mini-International Neuropsychiatric Interview (M.I.N.I.): the development and validation of a structured diagnostic psychiatric interview for DSM-IV and ICD-10. J Clin Psychiatry 59: Suppl 2022–33;quiz 34–57.
46. Schmand B, Bakker D, Saan R, Louman J (1991) [The Dutch Reading Test for Adults: a measure of premorbid intelligence level]. Tijdschrift voor Gerontologie en Geriatrie 22: 15–19.
47. Raven JC, Raven J, Court JH (1993) Manual for Raven's Progressive Matrices and Vocabulary Scales. Oxford: Oxford Psychologist Press.
48. Lezak MD (1995) Neuropsychological assessment (third ed.). New York: Oxford University Press.
49. Russell TA, Reynaud E, Herba C, Morris R, Corcoran R (2006) Do you see what I see? Interpretations of intentional movement in schizophrenia. Schizophr Res 81: 101–111.
50. Dickerson F, Boronow JJ, Ringel N, Parente F (1999) Social functioning and neurocognitive deficits in outpatients with schizophrenia: A 2-year follow-up. Schizophr Res 37: 13–20.
51. Van Der Does AJW, Dingemans PMAJ, Linszen DH, Nugter MA, Scholte WF (1996) Symptoms, cognitive and social functioning in recent-onset schizophrenia: A longitudinal study. Schizophr Res 19: 61–71.
52. Dickerson F, Boronow JJ, Ringel N, Parente F (1996) Neurocognitive deficits and social functioning in outpatients with schizophrenia. Schizophr Res 21: 75–83.
53. Martin F, Baudouin JY, Tiberghien G, Franck N (2005) Processing emotional expression and facial identity in schizophrenia. Psychiatry Res 134: 43–53.
54. Schneider F, Gur RC, Gur RE, Shtasel DL (1995) Emotional processing in schizophrenia: neurobehavioral probes in relation to psychopathology. Schizophr Res 17: 67–75.
55. Mandal MA, Page KM (1998) Facial expressions of emotions and schizophrenia: A review. Schizophr Bull 24: 399–412.
56. Kohler CG, Bilker W, Hagendoorn M, Gur RE, Gur RC (2000) Emotion recognition deficit in schizophrenia: association with symptomatology and cognition. Biol Psychiatry 48: 127–136.

57. Corcoran R, Mercer G, Frith CD (1995) Schizophrenia, symptomatology and social inference: investigating "theory of mind" in people with schizophrenia. Schizophr Res 17: 5–13.
58. Mandal MK, Jain A, Haque-Nizamie S, Weiss U, Schneider F (1999) Generality and specificity of emotion-recognition deficit in schizophrenic patients with positive and negative symptoms. Psychiatry Res 87: 39–46.
59. Leitman DI, Foxe JJ, Butler PD, Saperstein A, Revheim N, et al. (2005) Sensory contributions to impaired prosodic processing in schizophrenia. Biol Psychiatry 58: 56–61.
60. Aleman A, Kahn RS, Selten JP (2003) Sex differences in the risk of schizophrenia: evidence from meta-analysis. Arch Gen Psychiatry 60: 565–571.
61. Takahashi H, Koeda M, Oda K, Matsuda T, Matsushima E, et al. (2004) An fMRI study of differential neural response to affective pictures in schizophrenia. Neuroimage 22: 1247–1254.
62. Rajarethinam RP, DeQuardo JR, Nalepa R, Tandon R (2000) Superior temporal gyrus in schizophrenia: A volumetric magnetic resonance imaging study. Schizophr Res 41: 303–312.
63. Dolan RJ, Fletcher P, Frith D, Friston KJ, Frackowiak RSJ, et al. (1995) Dopaminergic modulation of impaired cognitive activation in the anterior cingulate cortex in schizophrenia. Nature 378: 180–182.
64. Fletcher P, McKenna PJ, Friston KJ, Frith CD, Dolan RJ (1999) Abnormal cingulate modulation of fronto-temporal connectivity in schizophrenia. NeuroImage 9: 337–342.
65. Ashton L, Barnes A, Livingston M, Wyper D (2000) Cingulate abnormalities associated with PANSS negative scores in first episode schizophrenia. Behavioural Neurology 12: 93–101.
66. Shenton ME, Dickey CC, Frumin M, McCarley RW (2001) A review of MRI findings in schizophrenia. Schizophr Res 49: 1–52.
67. Mitelman SA, Shihabuddin L, Brickman AM, Hazlett EA, Buchsbaum MS (2005) Volume of the cingulate and outcome in schizophrenia. Schizophr Res 72: 91–108.
68. Shen D, Liu D, Liu H, Clasen L, Giedd J, et al. (2004) Automated morphometric study of brain variation in XXY males. NeuroImage 23: 648–653.
69. Bölte S, Holtmann M, Poustka F, Scheurich A, Schmidt L (2007) Gestalt Perception and Local-Global Processing in High-Functioning Autism. J Autism Dev Disord 37: 1493–1504.

70. Dakin S, Carlin P, Hemsley D (2005) Weak suppression of visual context in chronic schizophrenia. Curr Biol 15: R822–R824.

71. Uhlhaas PJ, Silverstein SM, Phillips WA, Lovell PG (2004) Evidence for impaired visual context processing in schizotypy with thought disorder. Schizophr Res 68: 249–260.

72. Bojesen A, Juul S, Gravholt CH (2003) Prenatal and postnatal prevalence of Klinefelter syndrome: a national registry study. J Clin Endocrinol Metab 88: 622–626.

The Moderating Role of Parental Smoking on their Children's Attitudes Toward Smoking among a Predominantly Minority Sample: A Cross-Sectional Analysis

Anna V. Wilkinson, Sanjay Shete and Alexander V. Prokhorov

ABSTRACT

Background

In general having a parent who smokes or smoked is a strong and consistent predictor of smoking initiation among their children while authoritative parenting style, open communication that demonstrates mutual respect between child and parent, and parental expectations not to smoke are protective. It has been hypothesized that parental smoking affects their children's smoking initiation through both imitation of the behavior and effects on attitudes toward smoking. The goals of the current analysis were to examine these two potential mechanisms.

Methods

In 2003, 1,417 high school students in Houston, Texas, completed a cross-sectional survey as part of the evaluation of an interactive smoking prevention and cessation program delivered via CD-ROM. To assess the relationship between number of parents who currently smoke and children's smoking status, we completed an unconditional logistic regression. To determine whether the attitudes that children of smokers hold toward smoking are significantly more positive than the attitudes of children of non-smokers we examined whether the parents smoking status moderated the relationship between children's attitudes toward smoking and their ever smoking using unconditional logistic regressions.

Results

Compared to participants whose parents did not currently smoke, participants who reported one or both parents currently smoke, had increased odds of ever smoking (OR = 1.31; 95% CI: 1.03–1.68; Wald $\chi^2 = 4.78$ *(df = 1)* $p = 0.03$ *and OR = 2.16; 95% CI: 1.51–3.10; Wald* $\chi^2 = 17.80$ *(df = 1)* $p < 0.001$*, respectively). In addition, the relationship between attitudes and ever smoking was stronger among participants when at least one parent currently smokes (OR = 2.50; 95% CI: 1.96–3.19; Wald* $\chi^2 = 54.71$ *(df = 1)* $p < 0.001$*) than among participants whose parents did not smoke (OR = 1.72; 95% CI: 1.40–2.12; Wald* $\chi^2 = 26.45$ *(df = 1)* $p < 0.001$*).*

Conclusion

Children of smokers were more likely to smoke and reported more favorable attitudes toward smoking compared to children of non-smokers. One interpretation of our findings is that parental smoking not only directly influences behavior; it also moderates their children's attitudes towards smoking and thereby impacts their children's behavior. Our results demonstrate a continued need for primary prevention smoking interventions to be sensitive to the family context. They also underscore the importance of discussing parental smoking as a risk factor for smoking initiation, regardless of ethnicity, and of tailoring prevention messages to account for the influence that parental smoking status may have on the smoking attitudes and the associated normative beliefs.

Introduction

Studies from the 1970s onwards have demonstrated that parental smoking and parental attitudes toward smoking are associated with smoking initiation among youth [e.g. [1-3]]. Flay et al. [4] reported that having a parent who smokes affects smoking initiation through imitation of the behavior and it also influences smoking attitudes, norms, and beliefs. Results from more recently published studies lend support to this claim. Having a parent who smokes or smoked is a strong and consistent predictor of smoking initiation among children [5-10], while authoritative parenting style [11,12], open communication that demonstrates mutual respect between child and parent [13,14], parental expectations not to smoke [15], and parental control [16] are protective.

To our knowledge, few, if any studies, have examined whether the strength of the reported association between attitudes toward smoking and ever smoking among children of smokers is significantly different from the strength of the association among children of non-smokers in a mostly minority sample. If children of smokers report significantly more positive attitudes toward smoking than children of non-smokers, this would suggest that having a parent who smokes modifies the relationship between attitudes and smoking, and provides evidence for an indirect effect of parental smoking on their children's behavior. In other words, parental smoking influences their children's attitudes toward smoking, which in turn increases the likelihood of the child smoking. This would suggest that interventions aimed at preventing adolescent smoking need to be sensitive to the family context and in particular underscores the need to discuss parental smoking with adolescents, especially with children of smokers.

Therefore our first goal was to examine the relationship between number of parents who currently smoked and children's ever smoking. We hypothesized that as the number of parents who currently smoke increases, so does the odds of their children's having ever smoked. Our second goal was to examine whether parental smoking modifies the relationship between children's attitudes toward smoking and children's ever smoking. We hypothesized that the relationship between attitudes and ever smoking is stronger when at least one parent is a current smoker.

Methods

This study presents a secondary analysis of baseline cross-sectional data collected as part of the evaluation of A Smoking Prevention Interactive Experience (ASPIRE). ASPIRE is an interactive smoking prevention and cessation program delivered via CD-ROM that has been implemented and evaluated in eight high schools in Houston, Texas. For a detailed description of the study design, recruitment procedures, and instrumentation, see [17]. Participants were in grade ten when they completed the evaluation between September 2002 and January 2003. Active consent was obtained from all participants; parents were asked to return a signed informed consent and students who were 18 and over signed their own consent forms. All aspects of this study received approval from the Institutional Review Board at The University of Texas M.D. Anderson Cancer Center, the Committee for the Protection of Human Subjects at The University of Texas at Houston School of Public Health and from Houston Independent School District's Research Department.

Participants

Participants included N = 1,417 or 88.6%, of the N = 1,599 high-school students in Houston who took part in the evaluation of ASPIRE. Of the 1,599 available for analysis 108 participants were missing data on the attitude measure and 74 were missing data on at least one of the parent variables.

Dependent Variable

The dependent variable in our study was ever smoking. We compared ever smokers (participants who had ever experimented with cigarettes) to never smokers, rather than simply comparing current experimenters to never smokers for two reasons. First, because we wanted to know if parental smoking influenced the odds of the child smoking, even just a puff of a cigarette, because 50% of children who even try a cigarette become smokers as adults [18]. And second, because smoking behavior at this age is quite volatile, it remains difficult to determine which child experimenter will become an adult smoker.

Parental Smoking Effects

To examine the relationship between number of parents who currently smoked and children's ever smoking we classified parental smoking status as "neither smokes," "only one smokes," or "both smoke." To examine whether parental smoking modifies the relationship between children's attitudes toward smoking and children's ever smoking, we determined if there were differences in the strength of the relationship between children's attitudes toward smoking and ever smoking by parental smoking status. In other words we posed the question, are the attitudes that children of smokers hold toward smoking significantly more positive than the attitudes of children of non-smokers? We assessed participants' attitudes using the Temptations to Smoke Scale, which is composed of 10 items and is tailored to the participant's smoking status [19]. Current and former smokers answer identical versions of the scale, while never smokers answer a modified version [20]. Using only items that were common across both versions, we created a four-item scale; two items assess negative affect (e.g. "When things are not going my way and I'm frustrated"), and two items assess functional aspects of smoking (e.g. "When I want to get thinner"). The revised scale had very good internal reliability (Cronbach's alpha = 0.79) and was treated as both a continuous and binary variable. As a binary variable we compared children who reported no temptations to children who reported some temptations.

Control Variables

We controlled for the participants' gender, age (examined as a continuous variable), and ethnicity (coded as "Black," "Hispanic," "White," or "Other"). The "Hispanic" group served as the reference category. We also controlled for parents' current marital status (coded as "married" or "not married") and the highest level of educational attainment of either parent (coded as "college degree or more" or "less than college degree"). Parents who were "not married" and those with "less than college degree" served as the reference categories, respectively. "Not married" parents included the following categories: separated, divorced, single, and widowed.

Statistical Analyses

Bivariate associations between participant smoking status and participant gender, participant ethnicity, parents' highest level of educational attainment, parents' marital status, and parents' smoking status were assessed using chi-squares. For participant ethnicity we created a dummy variable for each ethnic group and conducted two by two chi-square analyses. We took the same approach with parents' educational attainment. Mean differences in age and children's smoking attitudes by participant smoking status were assessed using Student's t-tests.

To assess the relationship between number of parents who currently smoked (neither, only one, or both) and children's smoking status (ever vs. never), we completed an unconditional logistic regression, adjusting for the control variables.

To assess whether parental smoking modifies the relationship between children's attitudes toward smoking and children's ever smoking, we followed a methodology outlined by Baron and Kenny [21] to determine whether the attitudes that children of smokers hold toward smoking are significantly more positive than the attitudes of children of non-smokers. In other words we examined whether the parents smoking status moderated the participants' attitudes toward smoking on their ever smoking. Although moderation is best determined prospectively, the presence of a moderator effect based on cross-sectional data will lend support to our hypothesis, underscoring the need to confirm the effect using longitudinal data. For this analysis, we collapsed parental smoking status into a two-level variable (neither parent smokes vs. either parent smokes). Next, we created an interaction term between parental smoking status and children's smoking attitudes. All three variables were simultaneously entered into an unconditional logistic regression model, adjusting for the control variables. To further assess the interaction effect, we created a four-level categorical composite variable of parental smoking and attitudes. Categories included: a) neither parents smoke, child reports no temptations; b) neither parents smoke, child reports some temptations; c) a parent smokes, child reports no temptations; and d) a parent smokes, child reports some temptations. Next we completed an unconditional logistic regression, adjusting for the control variables again.

Having established moderation (presence of a significant interaction term), we stratified the sample on parental smoking status and completed two unconditional logistic regressions. Both models included children's smoking attitudes and adjusted for the control variables.

Results

The participants included 569 males (40.1%) and 848 females (59.8%). The demographic characteristics of the participants and their parents by the participants' smoking status are presented in Table 1. The mean age of the participants was 15.66 years (SD = 0.90), and 587 (41.1%) reported ever smoking, of whom 17% reported smoking at least once every two weeks or more. Roughly 51% self-identified as Hispanic, 39% as black, and 6% as white; the remaining participants self-identified as Asian, American Indian, Alaska Native, Native Hawaiian, or Pacific Islander. The majority reported that their parents were married (52.3%) and that neither parent was a current smoker (55.9%); 32.3% reported that one parent currently smoked, and 11.8% reported that both parents currently smoked.

Results from the multivariable unconditional logistic regression examining the influence of the number of parents who currently smoked on children's ever smoking are presented in Table 2. Compared to their peers who reported that neither parent currently smoked, participants who reported that one parent currently smoked were 1.31 times (95% confidence interval [CI] = 1.03–1.68; Wald χ^2 = 4.78 (df = 1) $p = 0.03$) as likely to have ever smoked, and participants who reported that both parents currently smoked were 2.16 times (95% CI = 1.51–3.10; Wald χ^2 = 17.80 (df = 1) $p < 0.001$) as likely to have ever smoked. In addition, the overall p-value for number of parents who smoke was significant (p for trend < 0.001). Ever smoking also was associated with being male and older, living with parents' who highest level of education was less than a high school degree, while being black and living with parents who are married were protective.

Table 1. Demographic characteristics of participants and their parents by participants' smoking status (N = 1,417)*

	Participants' smoking status			
	Never (n = 830; 58.6%)	Ever (n = 587; 41.4%)	DF	p value
Participant Characteristic				
Gender				
Male	297 (52.2)	272 (47.8)		
Female	533 (62.9)	315 (37.1)	1	< 0.01
Age (years)				
Mean (SD)	15.60 (0.86)	15.74 (0.94)	1415	< 0.01
Ethnicity				
Black	375 (68.6)	172 (31.4)	1	< 0.01
Hispanic	376 (52.0)	347 (48.0)	1	< 0.01
Other	39 (62.9)	23 (37.1)	1	0.48
White	40 (47.1)	45 (52.9)	1	0.03
Temptations to smoke				
Mean (SD)	1.27 (0.65)	1.73 (0.90)	1415	< 0.01
Parent Characteristic				
Highest level of education of either parent				
Less than high school	136 (48.1)	147 (51.9)	1	< 0.01
Completed high school	188 (58.9)	131 (41.9)	1	0.88
Some college	157 (60.2)	104 (39.8)	1	0.57
Completed college	198 (63.3)	115 (36.7)	1	0.06
Marital status				
Married	463 (62.5)	278 (37.5)		
Not married	367 (54.8)	309 (45.2)	1	< 0.01
Smoking status				
Neither parent smokes	496 (63.6)	296 (36.4)		
Only one parent smokes	260 (56.8)	198 (43.2)		
Both parents smoke	74 (44.3)	93 (55.7)	2	< 0.01

*Data in table are numbers of participants or parents (percentages) unless otherwise specified; percentages are row percentages.
SD = Standard Deviation; DF = Degrees of Freedom.
P-values for age and temptations to smoke are based on Student's t-test, all other p-values are based on chi-square tests of association.

Table 2. Unconditional logistic regression examining the influence of number of parents who smoke on children's ever smoking (N = 1,417)

Characteristic	OR	95% CI		p value
Male	1.63	1.30	2.05	< 0.01
Age in years	1.14	1.01	1.29	0.04
Black vs. Hispanic	0.44	0.34	0.57	0.00
Other vs. Hispanic	0.68	0.39	1.18	0.17
White vs. Hispanic	1.05	0.65	1.69	0.86
Some HS vs. completed college	1.65	1.21	2.26	0.01
Completed HS vs. completed college	1.21	0.90	1.63	0.21
Some college vs. completed college	1.33	0.96	1.84	0.08
Parents are married	0.59	0.47	0.75	< 0.01
Only one parent smokes	1.31	1.03	1.68	0.03
Both parents smoke	2.16	1.51	3.10	< 0.01

OR, odds ratio; CI, confidence interval; ORs are adjusted for all other variables presented in the table. P-values are based on Wald chi-square with 1 df.

To determine if the attitudes that children of smokers hold toward smoking are significantly more positive than the attitudes of children of non-smokers (i.e. whether parental smoking status potentially moderates participants' attitudes toward smoking), we first examined the interaction term between parental smoking status and children's smoking attitudes (Baron and Kenny, 1986). After controlling for the participants' gender, age, and ethnicity, the parents' educational attainment and marital status, and the two main effects (children's attitudes and parental smoking status) the interaction term was significant (OR = 1.27; 95% CI = 1.07–1.52; Wald $\chi^2 = 7.41$ (df = 1) $p < 0.01$; data not shown). In other words, having at least one parent who currently smoked potentially does moderate the influence of children's smoking attitudes on ever smoking.

Results from the multivariable unconditional logistic regression examining the four-level categorical variable are presented in table 3. Compared to children whose parents do not smoke and who reported no temptations to smoke, their peers who reported some temptations were no more of less likely to be ever smokers; however children who live with at least one parent who smokes were more likely to be ever smokers. Among children who reported no temptations, the odds of ever smoking were 2.89 (95% CI = 2.09–3.97; Wald $\chi^2 = 41.49$ (df = 1) $p < 0.01$), but among the children who reported temptations, the odds of ever smoking were 5.26 (95% CI = 3.74–7.40; Wald $\chi^2 = 91.42$ (df = 1) $p < 0.01$). Please see figure 1 for the predicted probabilities of smoking based on the four-level categorical variable.

Table 3. Unconditional logistic regression examining the interaction between having a parent who smokes and child's attitudes toward smoking on children's ever smoking (N = 1,417)

Characteristic	OR	95% CI		*p* value
Neither parent smokes, child reports no temptations	1.00			
Neither parent smokes, child reports some temptations	1.16	0.864	1.56	0.32
A parent smokes, child reports no temptations	2.89	2.09	3.97	< 0.01
A parent smokes, child reports some temptations	5.26	3.74	7.40	< 0.01

OR, odds ratio; CI, confidence interval; ORs are adjusted for all other variables presented in the table. P-values are based on Wald chi-square with 1 df.
Control variables include child's age, gender, ethnicity and parents' educational attainment and marital status.

Next, we stratified the sample by parental smoking status and conducted multivariable logistic regression analyses to investigate the relationship between children's smoking attitudes and ever smoking (Table 4). Among participants whose parents were current nonsmokers, children's smoking attitudes were associated with a 1.72 times (95% CI = 1.40–2.12; Wald $\chi^2 = 26.45$ (df = 1) $p < 0.001$) increased risk of being an ever smoker, whereas among participants who had at least one parent who currently smoked, children's smoking attitudes were associated with a 2.50 times (95% CI = 1.96–3.19; Wald χ^2 = 54.71 (df = 1) p < 0.001) increased risk of being an ever smoker. Among participants whose parents were current nonsmokers, being male and older, as well having parents who have not completed high school were associated with ever smoking, while being black and having married parents were protective. Among participants with at least one parent who currently smoked, being male and having parents who have not completed high school were associated with ever smoking, while being black and parents being married were protective.

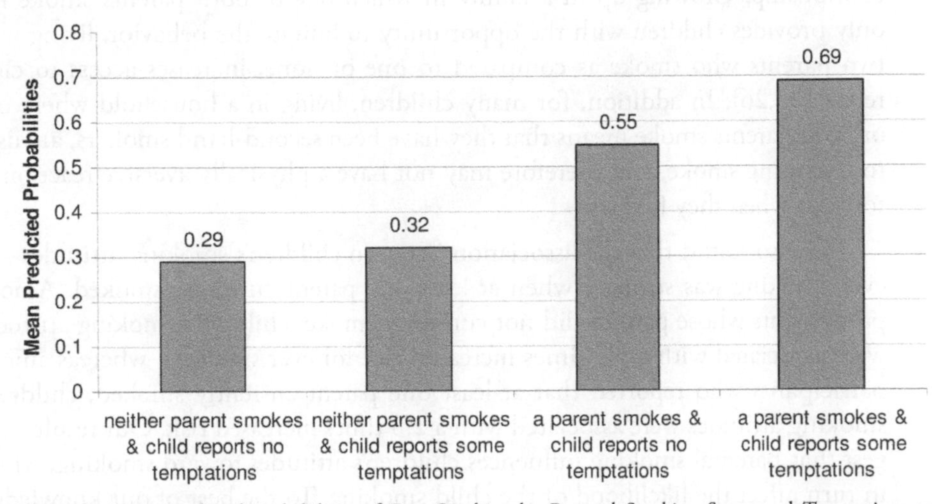

Figure 1. Mean Predicted Probabilities of Ever Smoking by Parental Smoking Status and Temptations to Smoke.

Table 4. Unconditional logistic regressions examining the influence of children's attitudes on their ever smoking, by parental smoking (N = 1,417)

	Neither parent smokes (*n* = 792)				At least one parent smokes (*n* = 625)			
Characteristic	**OR**	**95% CI**		***p* value**	**OR**	**95% CI**		***p* value**
Male	1.64	1.20	2.25	< 0.01	1.72	1.20	2.47	< 0.01
Age in years	1.21	1.02	1.42	0.03	0.98	0.80	1.20	0.82
Black vs. Hispanic	0.42	0.29	0.62	0.00	0.57	0.38	0.87	< 0.01
Other vs. Hispanic	0.66	0.31	1.39	0.66	0.81	0.33	2.00	0.65
White vs. Hispanic	1.00	0.49	2.03	0.99	1.23	0.61	2.50	0.57
Some HS vs. completed college	1.53	1.00	2.35	0.05	2.22	1.34	3.68	< 0.01
Completed HS vs. completed college	1.33	0.88	2.00	0.18	1.38	0.87	2.19	0.18
Some college vs. completed college	1.45	0.92	2.29	0.14	1.45	0.88	2.38	0.15
Parents are married	0.62	0.45	0.86	< 0.01	0.69	0.48	1.00	0.05
Temptations to smoke	1.72	1.40	2.12	0.00	2.50	1.96	3.19	< 0.01

OR, odds ratio; CI, confidence interval; ORs are adjusted for all other variables presented in the table. P-values are based on Wald chi-square with 1 df.

Discussion

In this study, we explored two mechanisms through which parental smoking may influence their children's smoking behavior and found evidence for both mechanisms. Consistent with previous research [22-24], we found that the odds for smoking increased with the number of parents who currently smoked. Compared to participants whose parents did not currently smoke, participants who reported that one parent currently smoked had a 1.3 times increased risk for ever smoking, and those who reported that both parents currently smoked had a 2.2 times increased risk. The p for trend was significant, suggesting a dose-response relationship. Growing up in a family in which one or both parents smoke not only provides children with the opportunity to imitate the behavior, living with two parents who smoke as compared to one or none, increases access to cigarettes [25,26]. In addition, for many children, living in a household where one or both parents smoke means that they have been second-hand smokers, are used to breathing smoke, and therefore may not have a physically aversive reaction to tobacco when they first try.

We also found that the association between children's smoking attitudes and ever smoking was stronger when at least one parent currently smoked. Among participants whose parents did not currently smoke, children's smoking attitudes were associated with a 1.7 times increased risk for ever smoking, whereas among participants who reported that at least one parent currently smoked, children's smoking attitudes were associated with a 2.5 times increased risk. Our results suggest that parental smoking influences children's attitudes toward smoking, which in turn affect the likelihood of the child smoking. To the best of our knowledge, although one previous study examined whether parental smoking mediates (is an intermediate step in the casual pathway) the relationship between children's attitudes toward smoking and smoking behavior [4] and another examined the relationship between children's implicit and explicit attitudes toward smoking and parental smoking [27], no studies have examined if parental smoking status moderates (modifies) the relationship between children's attitudes toward smoking and their ever smoking. However, because moderation is best determined prospectively, our results lend support to the hypothesis and need to be confirmed using longitudinal data.

Our results differ from those of the Chassin et al. [27] study, which found that neither the children's implicit nor their explicit attitudes toward smoking were associated with parental smoking status. Our results, however, are consistent with, and extend the earlier findings of Flay et al. [4], who noted that parental smoking was associated with changes in adolescent's explicit attitudes. Not only did we observe an overall association between parental smoking and children's explicit attitudes, we found that the relationship between children's attitudes toward smoking and their ever smoking was weaker among children of parents who did not currently smoke compared to children whose parents currently smoke. Again, because moderation is best determined prospectively, our results lend support to the hypothesis that parental attitudes moderate this relationship.

It is of concern that the relationship between attitudes toward smoking and ever smoking is stronger when at least one parent smokes because positive attitudes toward smoking among youth are associated with increased susceptibility to smoking [28], and youth who have experimented with cigarettes or who currently smoke [27,29,30] hold significantly more positive attitudes toward smoking than do their nonsmoking peers. Moreover, Chassin et al. [27] found that mothers who smoke report significantly more positive implicit and explicit attitudes toward smoking than mothers who have quit or never smoked, underscoring the possibility that positive attitudes toward smoking are learned and sanctioned at home.

The participants in the Chassin et al. [27] were predominantly non-Hispanic white, as were the majority of the participants in Ridner's study [30]. However, the participants in the Chalela study were all Latino, while the sample in the Flay et al. study [4] was multi-ethnic. Although the goals of these four studies and ours were different and all used different analytic approaches, all studies examined the relationship between attitudes toward smoking and behavior and all reported consistent results: positive attitudes were associated with ever smoking. Taken as a whole, this suggests that consistent with the conclusions drawn from other studies based on multi-ethnic samples [9,10], predictors of smoking behavior may be universal.

While our study focused on the family context, other factors may impact the relationship between children's attitudes towards smoking and their ever-smoking. For example, many studies have documented the role that peer influence [31,32] and perceptions of peer norms [33] play on smoking initiation. While outside the scope of the current study, examining peer influence, both separate from and in conjunction with parental influence, would refine our understanding of the relationship between attitudes and smoking.

This study has some limitations. First, the analysis is based on self-reported cross-sectional survey, limiting our ability to draw causal conclusions and test for moderation. Second, while most contemporary approaches to assessing family structure tend to compare differences between single and two parent families without regard to marital status, the data collected in this study did not permit such a distinction. The response categories probing parental marital status included married, separated, divorced, single, and widowed. Therefore we examined the influence of reporting married parents, which may serve as a proxy for two parent households, compared to reporting that parents are separated, divorced, single, or widowed. In future research we intend to fully examine the relationship between household structure, living arrangements and number of parents who smoke. Third, we do not know how long the participants were exposed to parental smoking, which limits our ability to determine if there is a threshold of exposure required to influence children. Fourth, we did not ask the ever smokers where they obtained the cigarettes they smoked. Therefore we cannot determine if current parental smoking directly increases access, and we cannot control for its potential influence in the analysis. Fifth, active consent was required of all students to participate in this study; more girls than boys returned their consent form resulting in the differential participation rates. Finally, we did not examine the influence of exposure to a parent who quit smoking while the participant was growing up, which has been shown to increase the likelihood of smoking [24]. However, the net result from the lack of information about past parental smoking would have underestimated the effect sizes observed in our study (biased the results toward the null), suggesting our findings may have been more pronounced had we adjusted for past parental smoking.

Conclusion

In conclusion, our study found evidence for two potential mechanisms through which one current parental smoking may influence children's smoking behavior. Parental smoking influences children's smoking independent of impacting child attitudes as well as influencing their children's attitudes toward smoking, which in turn may increase the likelihood of the child smoking. In addition to the mechanisms identified in our analysis, others have demonstrated that current parental smoking is associated with increased access to cigarettes [25], while living in a two parent household is protective [31,33]. Our results refine our understanding of how the family context contributes to smoking initiation, thereby demonstrating a continued need for primary prevention smoking interventions to be sensitive to the family context. Specifically, our results underscore the importance of discussing parental smoking as a risk factor for smoking initiation, regardless of ethnicity, and of tailoring prevention messages to account for the influence that parental smoking status may have on the smoking attitudes and the associated normative beliefs held by children of smokers.

Authors' Contributions

AVW conceived the current analysis, completed the analysis, and led the writing. SS oversaw the analysis and provided critical revisions. AVP conceived the original study, interpreted the results, and provided critical revisions.

Acknowledgements

This research is supported by the National Cancer Institute grants CA81934-0182 (Alexander V. Prokhorov) and CA105203-01A1 (Anna V. Wilkinson).

References

1. Borland BL, Rudolph JP: Relative effects of low SES, parental smoking, and poor scholastic performance on smoking among high school students. Soc Sci Med 1975, 9:27–30.
2. Gottlieb NH: The effects of peer and parental smoking on the smoking careers of college women: A sex-related phenomenon. Sco Sci Med 1982, 16:595–600.
3. Newman IM, Ward JM: The influence of parental attitude and behavior on early adolescent cigarette smoking. J Sch Health 1989, 59:150–152.
4. Flay BR, Hu FB, Siddiqui O, Day LE, Hedeker D, Petraitis J, Richardson J, Sussman S: Differential influence of parental smoking and friends' smoking on adolescent initiation and escalation of smoking. J Health Soc Beh 1994, 35:248–265.
5. Bricker JB, Peterson AV, Anderson MR, Leroux BG, Rajan B, Sarason IG: Close friends', parents', and older siblings' smoking: Reevaluating their influence on children's smoking. Nicotine Tob Res 2006, 8:217–226.
6. Bricker JB, Peterson AV, Leroux BG, Anderson MR, Rajan KB, Sarason IG: Prospective prediction of children's smoking transitions: role of parents' and older siblings' smoking. Addiction 2006, 101:128–136.
7. Chassin L, Presson CC, Pitts SC, Sherman SJ: The natural history of cigarette smoking from adolescence to adulthood in a Midwestern community sample: Multiple trajectories and their psychosocial correlates. Health Psychol 2000, 19:223–231.
8. Fleming CB, Hyoshin K, Harachi TW, Catalano RF: Family processes for children in early elementary school as predictors of smoking initiation. J Adolesc Health 2002, 30:184–189.
9. Griesler PC, Kandel DB, Cavies M: Ethnic differences in predictors of initiation and persistence of adolescent cigarette smoking in the National Longitudinal Survey of Youth. Nicotine Tob Res 2002, 4:79–93.
10. Kandel DB, Gedre-Egziabher K, Schaffran C, Hu MC: Racial/ethnic differences in cigarette smoking initiation and progression to daily smoking: a multilevel analysis. Am J Public Health 2004, 94:128–135.

11. Hill KG, Hawkins JD, Catalano RF, Abbott RD, Guo J: Family influences on the risk of daily smoking initiation. J Adolesc Health 2005, 37:202–210.
12. Simmons-Morton BG, Crump AD, Haynie DL, Saylor KE, Eitel P, Yu K: Psychosocial School and parent factors associated with recent smoking among early adolescent boys and girls. Prev Med 1999, 28:138–148.
13. Distefan JM, Gilpin EA, Choi WS, Pierce JP: Parental influences predict adolescent smoking in the United States, 1989–1993. J Adolesc Health 1998, 22:446–474.
14. Otten R, Harakeh Z, Vermulst AA, Eijnden RJJM, Engels RCME: Frequency and quality of parental communications as antecedents of adolescent smoking cognitions and smoking onset. Psychol Addict Behav 2007, 21:1–12.
15. Simons-Morton BG: The protective effect of parental expectations against early adolescent smoking initiation. Health Educ Res 2004, 19:561–569.
16. Blokland EA, Hale WW 3rd, Meeus W, Engels RC: Parental support and control and early adolescent smoking: a longitudinal study. Substance Use & Misuse 2007, 42:2223–32.
17. Prokhorov AV: Interventions for high-school and college students guided by the Transtheoretical Model of Change. In Research on the Transtheoretical Model: Where are we now, where are we going?. Edited by: Keller S, Velicer WF. Lengerich, Germany: Pabst Science Publishers; 2004:94–96.
18. U.S. Department of Health and Human Services: Preventing Tobacco use Among Young People: A Report of the Surgeon General. Atlanta Georgia: U.S. Department of Health and Human Services, Pubic Health Service, Centers for Disease Control and Prevention, National Center for Chronic Disease Prevention and Health Promotion, Office on Smoking and Health; 1994.
19. Pallonen UE: Transtheoretical measures for adolescent and adults smokers: similarities and differences. Prev Med 1998, 27:A29–A38.
20. Hudmon KS, Prokhorov AV, Koehly LM, DiClemente CC, Gritz ER: Psychometric properties of the decisional balance scale and the temptation to try smoking inventory in adolescents. J Child Adolesc Substance Abuse 1998, 6:1–18. (1997)
21. Baron RM, Kenny DA: The moderator-mediator variable distinction in social psychological research: Conceptual, strategic and statistical considerations. J Pers Soc Psychol 1986, 51:1173–1182.
22. Farkas AJ, Distefan JM, Choi WS, Gilpin EA, Pierce JP: Does Parental Smoking Cessation Discourage Adolescent Smoking? Prev Med 1999, 28:213–218.

23. Patterson AV, Leroux BG, Bricker J, Kealey KA, Marek PM, Sarason IG, Anderson MR: Nine-year prediction of adolescent smoking by number of smoking parents. Addict Behav 2006, 31:788–801.

24. Otten R, Engels RCME, Ven MOM, Bricker JB: Parental smoking and adolescent smoking stages: The role of parents' current and former smoking and family structure. J Behav Med 2007, 30:143–154.

25. DiFranza JR, Coleman M: Sources of tobacco for youths in communities with strong enforcement of youth access laws. Tobacco Control 2001, 10:323–328.

26. Jackson C, Henriksen L, Dickinson D, Messer L, Robertson SB: A longitudinal study predicting patterns of cigarette smoking in late childhood. Health Educ Behav 1998, 25:436–47.

27. Chassin L, Presson C, Rose J, Sherman SJ, Prost J: Parental smoking cessation and adolescent smoking. J Pediat Psychol 2002, 27:485–496.

28. Leatherdale ST, Brown KS, Cameron R, McDonald PW: Social modeling in the school environment, student characteristics, and smoking susceptibility: a multi-level analysis. J Adolesc Health 2005, 37:330–336.

29. Chalela P, Velez LF, Ramirez AG: Social influences, and attitudes and beliefs associated with smoking among border Latino Youth. J School Health 2007, 77:187–95.

30. Ridner SL: Predicting smoking status in a college-age population. Public Health Nurs 2005, 22:494–505.

31. Gritz ER, Prokhorov AV, Hudmon KS, Chamberlain RM, Taylor WC, DiClemente CC, Johnston DA, Hu S, Jones L, Mullin Jones M, Rosenblum CK, Ayars CL, Amos CI: Cigarette smoking in a multiethnic population of youth: methods and baseline findings. Prev Med 1998, 27:365–384.

32. Gritz ER, Prokhorov AV, Hudmon KS, Jones MM, Rosenblum C, Chang CC, Chamberlain RM, Taylor WC, Johnston D, de Moor C: Predictors of susceptibility to smoking and ever smoking: a longitudinal study in a triethnic sample of adolescents. Nicotine Tob Res 2003, 5:493–506.

33. Forrester K, Biglan A, Severson HH, Smolkowski K: Predictors of smoking onset over two years. Nicotine Tob Res 2007, 9:1259–1267.

Women's Childhood and Adult Adverse Experiences, Mental Health, and Binge Drinking: The California Women's Health Survey

Christine Timko, Anne Sutkowi, Joanne Pavao and Rachel Kimerling

ABSTRACT

Background

This study examined sociodemographic, physical and mental health, and adult and childhood adverse experiences associated with binge drinking in a representative sample of women in the State of California.

Materials and Methods

Data were from the 2003 to 2004 (response rates of 72% and 74%, respectively) California Women's Health Survey (CWHS), a population-based, random-digit-dial annual probability survey sponsored by the California Department of Health Services. The sample was 6,942 women aged 18 years or older.

Results

The prevalence of binge drinking was 9.3%. Poor physical health, and poorer mental health (i.e., symptoms of PTSD, anxiety, and depression, feeling overwhelmed by stress), were associated with binge drinking when demographics were controlled, as were adverse experiences in adulthood (intimate partner violence, having been physically or sexually assaulted, or having experienced the death of someone close) and in childhood (living with someone abusing substances or mentally ill, or with a mother vicimized by violence, or having been physically or sexually assaulted). When adult mental health and adverse experiences were also controlled, having lived as a child with someone who abused substances or was mentally ill was associated with binge drinking. Associations between childhood adverse experiences and binge drinking could not be explained by women's poorer mental health status in adulthood.

Conclusion

Identifying characteristics of women who engage in binge drinking is a key step in prevention and intervention efforts. Binge drinking programs should consider comprehensive approaches that address women's mental health symptoms as well as circumstances in the childhood home.

Background

Binge drinking has been linked to a broad range of health and psychosocial problems. Although definitions of binge drinking vary somewhat among studies, a commonly used definition for women and men is that of the Centers for Disease Control and Prevention's (CDC) Behavioral Risk Factor Surveillance System: five drinks per drinking occasion at least once during the past 30 days. Rates of binge drinking are lower among women than men across different age and racial/ethnic groups [1-3]. The rates of binge drinking among women have remained stable over the past two decades [4].

It is important to understand the determinants of binge drinking among women because higher rates of binge drinking are linked to increases in alcohol use disorders, injuries and other alcohol-related health problems (including poor health outcomes among babies born to alcohol abusing mothers), psychosocial problems, and high-risk behaviors (e.g., smoking, having multiple sexual partners) [5]. This study examined the rate of binge drinking and factors associated with binge drinking in a representative sample of women in California. We looked at factors in five domains in addition to sociodemographic characteristics: physical health, mental health, help with mental health problems, adverse adult experiences, and adverse childhood experiences. We also examined whether adverse childhood experiences were related to binge drinking as an adult, when adverse experiences and symptoms of poor mental health in adulthood were considered. The purpose was to indicate whether negative childhood experiences may predict binge drinking among adult women above and beyond more recent and current problems. Our goal is to help build binge-drinking prevention and intervention programs that are specific to women by targeting the factors associated with binge drinking.

Characteristics of Women Who Binge Drink

A number of studies have examined sociodemographic factors associated with women's binge drinking. Binge drinking is more common among younger (<30 years) women [6] and binge drinking as a younger woman increases the odds of binge drinking in middle age [7]. In addition, binge drinking is more likely among unmarried and less educated women [5], although studies of highly educated employees found binge drinking rates among women to be high [8,9]. Binge drinking was more common among White and mixed-race than among Hispanic, Black, or Asian women [6]. It is also known to be more common among American Indian women than women in other racial/ethnic groups [10,11]. Rates of binge drinking were higher among non-pregnant than pregnant women [6]. In the domain of mental health, higher rates of post-traumatic stress disorder (PTSD), anxiety, and depression were associated with more binge drinking [5]. In contrast, another study found that women with anxiety and depression had reduced odds of binge drinking [1].

Female victims of violence often engage in self-destructive and maladaptive coping behaviors, including binge drinking. Increased rates of domestic violence among women were associated with increased binge drinking rates [5]. Similarly, women's rates of binge drinking increased from 5.5% among those with no lifetime history of intimate partner violence (IPV), to 12.1% among those with a low level of IPV, to 16.8% among those with a moderate or high level [12]. Binge drinking may occur as a form of self-medication to alleviate symptoms of trauma, anxiety, and depression, and increase feelings of mastery and control [13].

Adverse experiences in childhood also increase the risk for the subsequent problematic use of alcohol such as binge drinking. Childhood sexual abuse was an important predictor of alcohol misuse and binge drinking among adult women [14,15], and adolescent victims of sexual and/or physical assault were more likely than controls to subsequently engage in binge drinking [13]. Each of eight kinds of adverse childhood experiences (verbal, physical, and sexual abuse, had a battered mother, lived with problem drinker/alcoholic or user of street drugs, mental illness in the household, parental separation or divorce, incarcerated household member) was associated with a higher risk of alcohol abuse in a sample of adult women and men [16].

Although adverse childhood experiences have been found to be associated with problematic alcohol use such as binge drinking, studies have not examined whether the childhood events are associated with alcohol misuse even when adult life events and circumstances are also considered. In addition, they have not examined whether poor mental health as an adult mediates the link between adverse childhood experiences and adult binge drinking, as some have hypothesized [17]. Both of these possibilities are important to explore because they have implications for the prevention and treatment of binge drinking. If adverse childhood experiences contribute to binge drinking when concurrent adult functioning is accounted for, and the association of adverse childhood experiences with binge drinking is not fully explained by its impact on poorer adult mental health, then specialized treatment for adverse childhood experiences must be integrated into clinical approaches used in preventing and treating binge drinking.

In summary, in this study we asked the following questions: (1) What is the rate of binge drinking in a representative sample of women in the State of California, (2) What are the sociodemographic, physical and mental health, and adverse adult and childhood experiences associated with binge drinking, (3) Are adverse childhood experiences associated with binge drinking, even when adverse adult experiences and mental health symptoms are already accounted for, and (4) Does poorer adult mental health mediate the association between adverse childhood experiences and adult binge drinking among women.

Materials and Methods

All procedures for this study were reviewed and approved by Stanford University's Institutional Review Board. We used data from the 2003 and 2004 California Women's Health Survey (CWHS), a population-based, random-digit-dial (not including cell phones) annual probability survey sponsored by the California Department of Health Services and designed in collaboration with several other state agencies and departments. Interviews for the CWHS are conducted in English or Spanish and take approximately 30 minutes to complete.

The response rates for the surveys (i.e., the proportions of eligible households contacted that resulted in a completed interview) were 72% and 74%, respectively. The combined 2-year sample consisted of 6,942 women aged 18 years or older. More detailed information about the sample procedures can be found in [18].

Measures

Binge Drinking

Respondents were asked: Considering all types of alcoholic beverages, how many times during the past month did you have five or more drinks on an occasion? Following the CDC definition used prior to 2006, respondents who reported doing so one time or more were classified as binge drinkers; all other respondents, including abstainers, were classified as non-binge drinkers.

Sociodemographics

Respondents reported their age, race and ethnicity, marital status, whether children were in the household, education, and employment status. Respondents' reports of household size and annual household income were used to determine whether they were above or below the federal poverty line [19].

Physical Health

The interview asked: To your knowledge, are you now pregnant (yes/no). It also asked: Do you now smoke cigarettes. Answers were classified as yes (every day or some days) or no (not at all). Respondents' self-reported health was classified as good (i.e., good, very good, or excellent) or fair/poor.

Mental Health

Symptoms of PTSD were assessed using a five-item screen, demonstrated by Prins et al. [20] to detect current (past 30 days) and clinically significant PTSD symptoms. One item is a general trauma probe in which the interviewer asks if the respondent has ever, in her lifetime, had any experience or experiences that are frightening, horrible or upsetting. Four more items assess the presence or absence of the following PTSD symptoms: intrusive trauma-related thoughts, avoidance of trauma-related cues, emotional numbing, and physiological hyperarousal. Respondents were coded positive for PTSD symptoms if they screened positive for trauma and endorsed ≥3 PTSD items. To assess depression, respondents were asked to report how many of the past 30 days they felt sad, blue, or depressed; and to assess anxiety, they reported how many days they felt worried, tense, or anxious. These items were from the healthy days measure used in the Brief Behavioral Risk Factor Survey [21]. Respondents were coded positive for depression or anxiety if they reported symptoms on ≥14 days of the past 30 days.

Mental Health Help

Poor mental health was assessed by asking respondents to report how many days during the past 30 their mental health (e.g., problems with emotions) was not good. Respondents were coded positive for poor mental health if they reported that their mental health was not good on ≥14 days. In addition, respondents reported whether they had wanted or needed help with personal or family problems from a mental health professional in the past 12 months (yes/no), and whether they had obtained such help (yes/no).

Adverse Adult Experiences

CWHS questions on adverse adult and childhood experiences were from the Traumatic Stress Schedule, a validated self-report measure of traumatic events [22], and subjected to cognitive testing for use in telephone surveys. Respondents reported whether, during the past 12 months, they had experienced intimate partner violence. Specifically, if the respondent answered yes to one or more of 8 items (e.g., partner or former partner had thrown something at her; had pushed, grabbed, shoved, or slapped her), she was classified as having experienced intimate partner violence. Respondents also reported whether they experienced the following adverse experiences during their adulthood (age 18 or after): physical assault (i.e., was beaten up, slapped, punched, kicked, or attacked by a stranger or someone known), sexual assault (i.e., was forced into unwanted sexual activity), or death of a close friend or family member in an accident, homicide, or suicide.

Adverse Childhood Experiences

Respondents reported whether they experienced the following adverse experiences during their childhood (before age 18): was removed from home and placed in foster care; lived with someone who was a problem drinker or alcoholic or used street drugs; lived with someone who was depressed or mentally ill; mother or stepmother was treated violently; lived with someone who went to prison or jail; emotional abuse (i.e., adult in household often or very often swore at, insulted, put down, or made her afraid that she would be physically hurt); physical assault (i.e., was beaten up, slapped, punched, kicked, or attacked by a stranger or someone known); and sexual assault (i.e., was forced into unwanted sexual activity).

Analyses

The first set of logistic regression analyses examined associations of sociodemographic factors (each entered singly as an independent variable) with binge drinker status. Then, we conducted logistic regression analyses that controlled for the sociodemographic factors found to be significantly associated with binge drinking, and entered one factor (i.e., one variable) in the domain of physical health, mental health, mental health help, adverse adult experiences, or adverse childhood experiences as a predictor. A third set of logistic regressions examined associations of adverse childhood experiences with binge drinking when sociodemographic, adult mental health, and adverse adult experiences that were significantly associated with binge drinking were controlled; the purpose was to determine whether adverse childhood experiences were independently associated with binge drinking after more recent adverse experiences and mental health symptoms as adults were considered. Finally, we conducted a series of regression analyses to determine whether poorer adult mental health mediated between adverse childhood experiences and binge drinking.

Results

The prevalence of binge drinking was 9.3%. The other sample characteristics are presented in Table 1. We first compared binge drinkers (n = 647) to non-binge drinkers (n = 6,295) on demographics (Table 2). Because binge drinkers were younger and more likely to be unmarried and unemployed, and were somewhat more likely to have an income below the federal poverty line, we controlled for these variables in all subsequent analyses.

Logistic regressions predicting binge drinking from current physical and mental health, and help for mental health problems in the past year, found a significant association for each factor (Table 3). Specifically, binge drinking was associated with not having a current pregnancy, currently smoking, and being in fair or poor rather than good health. In the mental health domain, symptoms of PTSD, anxiety, and depression, as well as feeling overwhelmed by stress, were associated with binge drinking. Consistently, poor mental health, and perceiving the need for and utilizing mental health help in the past year, were associated with binge drinker status. As expected, binge drinking was associated with adverse experiences in adulthood (intimate partner violence, having been physically or sexually assaulted, or having experienced the death of someone close) and in childhood (living with someone abusing substances or mentally ill, or with a mother victimized by violence, and having been physically or sexually assaulted) (Table 4).

Table 1. Sample characteristics

Sociodemographics	% of full sample
Age (mean, SD)	32.6, 13.0
Minority/racial group member	50.2
Not married	45.0
No. children in household	49.6
No education beyond high school	37.4
Not employed	45.4
Income below federal poverty line	17.9
Physical health	
Not pregnant	94.7
Smokes	13.6
Fair/poor health	17.3
Mental health	
PTSD symptoms	6.3
Anxiety symptoms	21.8
Depression symptoms	13.7
Overwhelmed by stress	12.5
Mental health help	
Poor mental health	14.2
Perceived need for mental health help	23.2
Utilized mental health help	9.8
Adult victimization	
Intimate partner violence	9.9
Physically assaulted	19.7
Sexually assaulted	10.6
Death of someone close	29.7
Childhood victimization	
Lived in foster care	3.4
Lived with someone abusing substances	22.3
Lived with someone mentally ill	16.8
Lived with mother victimized by violence	18.4
Lived with someone incarcerated	7.6
Emotionally abused by adult in home	18. 4
Physically assaulted	19.1
Sexually assaulted	11.2

Table 2. Logistic regressions predicting binge (n = 647) versus non-binge (n = 6295) drinker status from sociodemographics

	Binge drinker			
Sociodemographics	B	OR	95% CI	p
Age (mean, SD)	-.477	.621	.58–.66	<.001
Minority racial/ethnic group member	.084	1.087	.93–1.28	.311
Not married	.980	2.664	2.25–3.16	<.001
No children in household	.135	1.144	.97–1.35	.104
No education beyond high school	.017	1.018	.86–1.20	.838
Not employed	.429	1.535	1.30–1.82	<.001
Income below federal poverty line	.180	1.197	.98–1.46	.081

Note: Each p-value is from a Wald X^2 (df = 1).

Table 3. Logistic regressions predicting binge (n = 647) or non-binge (n = 6295) drinker status from physical and mental health and mental health help, controlling for sociodemographics

	Binge drinker			
	B	OR	95% CI	p
Physical Health				
Not pregnant	2.288	9.859	3.40–28.59	<.001
Smokes	.985	2.679	2.21–3.25	<.001
Fair/Poor health	.287	1.333	1.10–1.76	.041
Mental health				
PTSD symptoms	.787	2.197	1.69–2.86	<.001
Anxiety symptoms	.478	1.614	1.34–1.94	<.001
Depression symptoms	.497	1.644	1.33–2.04	<.001
Overwhelmed by stress	.613	1.846	1.50–2.28	<.001
Mental health help				
Poor mental health	.484	1.622	1.31–2.00	<.001
Perceived need for mental health help, past year	.305	1.357	1.13–1.63	.001
Utilized mental health help, past year	.479	1.614	1.19–2.20	.002

Note: Each p-value is from a Wald X^2 (df = 1).

Table 4. Logistic regressions predicting binge (n = 647) or non-binge (n = 6295) drinker status from adult and childhood adverse experiences, controlling for sociodemographics

	Binge drinker			
	B	OR	95% CI	p
Adult adverse experiences				
Intimate partner violence	.542	1.719	1.37–2.15	<.001
Physically assaulted	.670	1.954	1.61–2.36	<.001
Sexually assaulted	.279	1.321	1.02–1.71	.035
Death of someone close	.641	1.898	1.60–2.25	<.001
Childhood adverse experiences				
Lived in foster care	.288	1.334	.90–1.99	.157
Lived with someone abusing substances	.480	1.616	1.35–1.94	<.001
Lived with someone mentally ill	.594	1.811	1.50–2.19	<.001
Lived with mother victimized by violence	.587	1.799	1.36–2.39	<.001
Lived with someone incarcerated	.191	1.210	.93–1.58	.161
Emotionally abused by adult in home	.421	1.523	1.26–1.85	<.001
Physically assaulted	.513	1.671	1.39–2.01	<.001
Sexually assaulted	.424	1.527	1.21–1.92	<.001

Note: Each p-value is from a Wald X^2 (df = 1).

Because, as already noted, binge drinking is more common in younger age groups, we examined whether age was a moderator of associations between physical and mental health, mental health help, and adult and childhood adverse experiences and binge drinking. Specifically, we conducted each logistic regression in Tables 3 and 4, adding the interaction of age by the predictor. The interactions of age by poor mental health (B = -.272, OR = .762, CI = .63–.72, Wald X2 = 9.14, df = 1, p = .003) and by having lived with someone mentally ill (B = -.163, OR-.850, CI = .73–.99, Wald X2 = 4.03, df = 1, p = .045) in predicting binge drinking were significant. There were positive associations of poor mental health and having lived with someone mentally ill with binge drinking in each age group except for the oldest group. That is, among women 65 years old or older only, poor mental health and having lived with someone mentally ill were associated with non-binge drinker status.

We examined whether adverse childhood experiences were still associated with binge drinking when mental health (PTSD, anxiety, depression, overwhelmed) and adverse experiences (intimate partner violence, physical assault, sexual assault, death of someone close) in adulthood were controlled (Table 5). When adulthood factors were also considered, the childhood experiences of having lived with someone who abused substances or who was mentally ill were associated with binge drinking. Specifically, these childhood experiences were associated with roughly double the likelihood of binge drinking. Binge drinking occurred in 14% of women who lived with an individual abusing substances, and in 8% of women who did not. And, binge drinking occurred in 16% of women who lived with an individual who was mentally ill, and in 8% of women who did not.

Table 5. Logistic regressions predicting binge (n = 647) or non-binge (n = 6295) drinker status from childhood adverse experiences, controlling for sociodemographics, mental health, and adult adverse experiences

	Binge drinker			
Childhood adverse experiences	B	OR	95% CI	p
Lived in foster care	.044	1.045	.69–1.58	.835
Lived with someone abusing substances	.241	1.273	1.05–1.55	.015
Lived with someone mentally ill	.359	1.431	1.17–1.76	.001
Lived with mother victimized by violence	-.050	.951	.70–1.29	.744
Lived with someone incarcerated	-.121	.886	.67–1.17	.396
Emotionally abused by adult in home	.073	1.075	.87–1.34	.514
Physically assaulted	.188	1.206	.98–1.49	.081
Sexually assaulted	.094	1.099	.85–1.42	.464

Note: Each p-value is from a Wald X^2 (df = 1).

Finally, we examined whether poor mental health as an adult mediated associations between adverse childhood experiences and adult binge drinking. First, we created two new dichotomous variables reflecting adverse childhood experiences: (1) lived with a person who abused substances and/or was mentally ill and/or a mother victimized by violence (50.3%) or not, and (2) was emotionally abused, physically assaulted, or sexually assaulted (33.7%) or not. Regressions showed that living with a disordered/victimized person predicted binge drinking, as did childhood abuse/assault (Table 6).

To examine mediation, first, poor mental health (the potential mediator) was the dependent variable in regressions using the new variables as predictors. Living with a disordered or victimized person was associated with poor mental health, as was having been abused or assaulted (Table 6). As already noted (Table 4), poor mental health was associated with binge drinking. A logistic regression predicting binge drinker status from both living with a disordered/victimized person and poor mental health found that both factors retained significant associations with binge drinking (Table 6). Finally, a logistic regression predicting binge drinker status from both childhood abuse/assault and poor mental health also found that both factors retained significant associations (Table 6). That is, adverse childhood experiences represented by living with others' or their own difficult life circumstances retained independent associations with women's binge drinking and was not mediated by poorer mental health status in adulthood.

Table 6. Logistic regressions to examine adult mental health as a mediator between childhood adverse experiences and binge drinking.

			Binge drinker	
Childhood experiences	B	OR	95% CI	p
Living with a disordered/victimized person	0.609	1.84	1.25–1.88	<0.001
Abuse/assault	0.578	1.78	1.47–2.06	<0.001
			Adult poor mental health	
Living with a disordered/victimized person	0.964	2.62	2.20–3.13	<0.001
Abuse/assault	1.028	2.80	2.43–3.22	<0.001
			Binge drinker	
Childhood living with a disordered/victimized person	0.380	1.46	1.19–1.80	<.001
Adult poor mental health	0.309	1.36	1.06–1.75	0.016
			Binge drinker	
Childhood abuse/assault	0.500	1.65	1.39–1.96	<0.001
Adult poor mental health	0.357	1.43	1.15–1.78	0.001

Note: Each p-value is from a Wald X^2 ($df = 1$).

Discussion

The CWHS found that the prevalence of binge drinking among adult women in California was 9.3%. This is similar to findings from national surveys of the US population in which 7.3% to 11.8% of women reported drinking five or more drinks on an occasion in the past month [2,23,24].

Compared to women who did not binge drink, binge drinking women were less likely to be pregnant, and more likely to smoke cigarettes and to describe themselves as only in fair or poor health. These results agree with those from other studies that pregnancy is associated with less binge drinking whereas smoking is associated with more binging on alcohol [25,26]. One survey found that 38% of young women reported smoking as the main reason for drinking [27]. Young women who both binge drank and smoked were especially likely to report depressive symptoms [28].

We found that about 13%–14% of the women in this sample had symptoms of depression or overwhelming stress, whereas symptoms of PTSD were less common and symptoms of anxiety were reported by 22% of women. Having symptoms of depression, stress, PTSD, and anxiety were each associated with binge drinking. The association between depression and alcohol misuse has been relatively well-documented in the literature. For example, analyses of the Canadian National Population Health Survey found an association of major depression and binge drinking among women [29]. In a sample of 121 alcohol-dependent women, 31% were diagnosed with depression, and 46% with anxiety, suggesting that women often alleviate stress through drinking [30]. Consistently, we found that women's self-reports of poor mental health, and having perceived the need for and utilized help for mental health problems in the past year, were associated with binge drinking.

As expected, binge drinking was associated with adverse experiences both in adulthood and in childhood. For adult victimization by intimate partner violence or sexual assault in particular, drinking is often a maladaptive means of coping with the traumatic aftermath [31-34]. When victimization by such traumatic events is chronic, binge drinking may be explained by PTSD symptoms [35,36]. That is, alcohol may be used to medicate PTSD-related sleep difficulties, hyperarousal, and other symptoms [37,38].

Victimizing childhood experiences have been consistently linked to problems in childhood that extend into adulthood [39-41]. For example, there was a strong association between adverse childhood experiences such as physical and sexual assault and drug abuse in young women [42]. In addition, longitudinal studies found that mothers' depressive symptoms and history of victimization predicted poorer behavioral outcomes among the children, and that the risk of child behavior problems increased with the number of areas B substance use, mental health, or domestic violence B in which the mother reported difficulties [43,44]. Although the CWHS did not assess respondents' recollections of their psychiatric symptoms in childhood, possibly, those whose mothers had psychiatric difficulties or victimizing experiences had more dysfunction in childhood that then persisted into adulthood. Unfortunately, increased rates of binge drinking associated with childhood and adult victimization may increase the risk for re-victimization [45].

Even after adulthood factors were considered, the childhood experiences of having lived with someone who abused substances (22.3% of the sample did so) or who was mentally ill (16.8% did so) were still associated with binge drinking. Problematic alcohol use is known to be influenced by problematic parental drinking and a family history of alcoholism [46]. While alcoholism has distinct biological and genetic influences [46-48], parental heavy-drinking norms and approval of, or lack of attention to, offspring drinking are also important correlates of risky drinking behaviors and poorer drinking outcomes [49-51]. The risk conferred by living with a dysfunctional adult suggests possible benefits of family-oriented medical and mental health care.

Whether providers are initially focused on the adult or the child, there is the potential to help disrupt intergenerational alcohol misuse by considering the entire family or at least the parent-offspring dyad.

Importantly, we found that adverse childhood experiences represented by living in the midst of others' or one's own difficult life circumstances retained independent associations with women's binge drinking and was not mediated by poorer mental health status in adulthood. Thus it may not always be enough to intervene only with binge drinking, or even with the depression and anxiety also associated with binge drinking. Rather, providers working with binge drinkers should consider asking whether adverse childhood experiences have occurred (in this sample, one-half of women lived with an ill or victimized person in their childhood home, and one-third were personally abused or assaulted as children), and then addressing any consequences of such adverse experiences if they have taken place. Research on trauma-focused therapy for women who were victimized in childhood that has shown promise for improving mental health symptoms [52-54] needs to be extended to alcohol and drug outcomes.

Limitations and Conclusion

The findings must be considered in light of the methods used. The definition of binge drinking was based on a quantitative cut point that did not consider individual factors that affect blood alcohol concentrations (e.g., weight, drinking history, other drugs ingested). Although a clinically-focused definition (e.g., feelings of intoxication) may have produced different results, consuming five or more drinks on a single occasion usually results in intoxication and impairment [55].

Self-reports as used here may somewhat underestimate alcohol consumption compared with diagnostic interviews, but they typically provide more accurate data than laboratory tests and collateral reports [56]. The data on women's health and binge drinking were cross-sectional, and so cause and effect cannot be determined from these results. Because data on adverse experiences as children and adults were retrospective, it is possible that recall bias influenced self-reports. However, previous research suggests that negative experiences in childhood tend to be under- rather than over-reported [57-59], perhaps leading to an underestimation of the strength of associations between childhood and adult hardship and binge drinking. An additional limitation is that the CWHS did not obtain diagnostic information with regard to physical health, PTSD, anxiety, and depression. Although CWHS items were selected from previously conducted national or statewide surveys for comparability, our findings need replication with more sensitive and precise measures of self-perceived physical and mental health.

Identifying characteristics that are commonly found among women who engage in binge drinking is a key step in designing and implementing effective prevention and intervention efforts. Binge drinking prevention programs perhaps should attend to adult mental health symptoms as well as dynamics in the childhood home. Experts working to decrease binge drinking may consider screening for addiction and other psychiatric disorders reported in the childhood home as well as other types of adverse childhood experiences and then addressing the trauma and other consequences of those stressors. Findings that childhood abuse is associated with poor substance use disorder treatment outcomes in adulthood [60] underscore the need to examine comprehensive and integrated approaches to prevent continued binge drinking.

Competing Interests

The authors declare that they have no competing interests.

Authors' Contributions

All the authors contributed to the design of the study and the analysis and interpretation of data, RK and JP helped to design and acquire data from the CWHS, CT managed the literature searches and wrote the first draft of the manuscript, AS managed the analyses. All authors contributed to and approved the final manuscript.

Acknowledgements

This project was funded by the VA Office of Research and Development (Health Services Research and Development Service, RCS 00-001). We thank Rudolf Moos for comments on an earlier draft. The opinions expressed here are the authors' and do not necessarily represent the views of the Department of Veterans Affairs.

References

1. Haynes JC, Farrell M, Singleton N, Meltzer H, Araya R, Lewis G, Wiles NJ: Alcohol consumption as a risk factor for anxiety and depression. Br J Psychiatry 2005, 187:544–551.

2. Naimi TS, Brewer RD, Mokdad A, Denny C, Serdula MK, Marks JS: Binge drinking among US adults. JAMA 2003, 289:70–75.
3. Wiscott R, Kopera-Frye K, Begovic A: Binge drinking in later life: Comparing young-old and old-old social drinkers. Psychol Addict Behav 2002, 16:252–255.
4. Mack KA, Ahluwalia IB: Monitoring women's health in the United States: Selected chronic disease indicators, 1991–2001 BRFSS. J Womens Health 2003, 12(4):309–314.
5. Bradley KA, Bush KR, Davis TM, Dobie DJ, Burman ML, Rutter CM, Kivlahan DR: Binge drinking among female Veterans Affairs patients: Prevalence and associated risks. Psychol Addict Behav 2001, 15:297–305.
6. Caetano R, Ramisetty-Mikler S, Floyd LR, McGrath C: The epidemiology of drinking among women of child-bearing age. Alcohol Clin Exp Res 2006, 30:1023–1030.
7. Jefferis BJMH, Power C, Manor O: Adolescent drinking level and adult binge drinking in a national birth cohort. Addiction 2005, 100:543–549.
8. Matano RA, Koopman C, Wanat SF, Whitsell S, Borggrefe A, Westrup D: Assessment of binge drinking of alcohol in highly educated employees. Addict Behav 2003, 28:1299–1310.
9. Matano RA, Wanat SF, Westrup D, Koopman C, Whitsell SD: Prevalence of alcohol and drug use in a highly educated workforce. J Behav Health Serv Res 2002, 29:30–44.
10. Bursac Z, Campbell JE: From risky behaviors to chronic outcomes: Current status and Healthy People 2010 goals for American Indians in Oklahoma. J Okla State Med Assoc 2003, 96:569–573.
11. Denny CH, Holtzman D, Cobb N: Surveillance for health behaviors of American Indians and Alaska Natives. MMWR Surveill Summ 2003, 52(7):1–13.
12. McNutt L, Carlson BE, Persaud M, Postmus J: Cumulative abuse experiences, physical health and health behaviors. Ann Epidemiol 2002, 12:123–130.
13. Kaukinen C: Adolescent victimization and problem drinking. Violence Vict 2002, 17:669–689.
14. Galaif ER, Stein JA, Newcomb MD, Bernstein DP: Gender differences in the prediction of problem alcohol use in adulthood: Exploring the influence of family factors and childhood maltreatment. J Stud Alcohol 2001, 62:486–493.

15. Jasinski JL, Williams LM, Siegel J: Childhood physical and sexual abuse as risk factors for heavy drinking among African-American women: A prospective study. Child Abuse Negl 2000, 24:1061–1071.

16. Dube SR, Anda RF, Felitti VJ, Edwards VJ, Croft JB: Adverse childhood experiences and personal alcohol abuse as an adult. Addict Behav 2002, 27:713–725.

17. Downs WR, Harrison L: Childhood maltreatment and the risk of substance problems in later life. Health Soc Care Community 1998, 6:35–46.

18. Kimerling R, Baumrind N: Intimate partner violence and use of welfare services among California women. J Sociol Soc Welf 2004, 31:161–176.

19. Federal Register: 2002 HHS poverty guidelines. Fed Regist 2002, 67:6931–6933.

20. Prins A, Ouimette P, Kimerling R, Cameron R, Hugelshofer D, Shaw-Hegwer J, Thrailkill A, Gusman FD, Sheikh JI: The Primary Care PTSD Screen: Development and operating characteristics. Prim Care Psychiatry 2004, 9:9–14.

21. Moriarty DJ, Zack MM, Kobau R: The Center for Disease Control and Prevention's Healthy Days Measures- Population tracking of perceived physical and mental health over time. Health Qual Life Outcomes 2003, 1:37.

22. Norris FH: Screening for traumatic stress: A scale for use in the general population. J Appl Soc Psychol 1990, 20:1704–1718.

23. Ebrahim SH, Diekman ST, Floyd RL, Decoufle P: Comparison of binge drinking among pregnant and nonpregnant women, United States, 1991–1995. Am J Obstet Gynecol 1999, 180:1–7.

24. Miller JW, Gfoerer JC, Brewer RD, Naimi TS, Mokdad A, Giles WH: Prevalence of adult binge drinking: A comparison of two national surveys. Am J Prev Med 2004, 27:197–204.

25. Strine TW, Okoro CA, Chapman DP, Balluz LS, Ford ES, Ajani UA, Mokdad AH: Health-related quality of life and health risk behavior among smokers. Am J Prev Med 2005, 28:182–187.

26. Tsai J, Floyd R, Green P, Boyle C: Patterns and average volume of alcohol use among women of childbearing age. Matern Child Health J 2007, 11:437–445.

27. Schoen C, Davis K, Collins K, Greenberg L, Des Roches C, Abrams M: The Commonwealth Fund survey of the health of adolescent girls. New York: Commonwealth Fund; 1997.

28. Pirkle EC, Richter L: Personality, attitudinal and behavioral risk profiles of young female binge drinkers and smokers. J Adolesc Health 2006, 38(1):44–54.
29. Wang J, Patten SB: Alcohol consumption and major depression: Findings from a follow-up study. Can J Psychiatry 2001, 46:632–638.
30. Dunne FJ, Galatopoulos C, Schipperheijn JM: Gender differences in psychiatric morbidity among alcohol misusers. Compr Psychiatry 1993, 34:95–101.
31. Kilpatrick DG, Acierno R, Resnick HS, Saunders BE, Best CL: A 2-year longitudinal analysis of the relationships between violent assault and substance use in women. J Consult Clin Psychol 1997, 65:834–847.
32. Miranda JJ, Vilchez E: So simple and so meaningful: An approach to mental health after violence. J Epidemiol Community Health 2002, 56:642.
33. Rice C, Mohr CD, Del Boca FK, Mattson ME, Young L, Brady K, Nickless C: Self-reports of physical, sexual and emotional abuse in an alcoholism treatment sample. J Stud Alcohol 2001, 62:114–123.
34. Testa M, Livingston JA, Leonard KE: Women's substance use and experiences of intimate partner violence: A longitudinal investigation among a community sample. Addict Behav 2003, 28:1649–1664.
35. Jacobsen LK, Southwick SM, Kosten TR: Substance use disorders in patients with posttraumatic stress disorder: A review of the literature. Am J Psychiatry 2001, 158:1184–1190.
36. Stewart SH: Alcohol abuse in individuals exposed to trauma: A critical review. Psychol Bull 1996, 120:83–112.
37. Nishith P, Resick PA, Mueser KT: Sleep difficulties and alcohol use motives in female rape victims with posttraumatic stress disorder. J Trauma Stress 2001, 14:469–479.
38. Stewart SH, Conrod PJ, Samoluk SB, Pihl RO, Dongier M: Posttraumatic stress disorder symptoms and situation-specific drinking in women substance abusers. Alcohol Treat Q 2000, 18:31–47.
39. Chapman DP, Whitfield CL, Felitti VJ, Dube SR, Edwards VJ, Anda RF: Adverse childhood experiences and the risk of depressive disorders in adulthood. J Affect Disord 2004, 82:217–225.
40. Kessler R, Davis C, Kendler K: Childhood adversity and adult psychiatric disorder in the US National Comorbidity Survey. Psychol Med 1997, 27:1101–1119.

41. Horwitz AV, Widom CS, McLaughlin J, White HR: The impact of childhood abuse and neglect on adult mental health: A prospective study. J Health Soc Behav 2001, 42:184–201.
42. Schilling EA, Aseltine RH, Gore S: Adverse childhood experiences and mental health in young adults: A longitudinal survey. BMC Public Health 2007, 7:30–51.
43. Thompson R: Mothers' violence victimization and child behavior problems: Examining the link. Am J Orthopsychiatry 2007, 77:306–315.
44. Whitaker RC, Orzol SM, Kahn RS: Maternal mental health, substance use, and domestic violence in the year after delivery and subsequent behavior problems in children at age 3 years. Arch Gen Psychiatry 2006, 63:551–560.
45. Gidycz CA, Loh C, Lobo T, Rich C, Lynn SJ, Pashdag J: Reciprocal relationships among alcohol use, risk perception, and sexual victimization: A prospective analysis. J Am Coll Health 2007, 56:5–14.
46. Schuckit MA: Biological, psychological and environmental predictors of the alcoholism risk: A longitudinal study. J Stud Alcohol 1998, 59:485–494.
47. Crabbe J: Alcohol and genetics: New trouble. Am J Med Genet 2002, 114:969–974.
48. Pickens R, Svikis D, McGue M, LaBuda M: Common genetic mechanisms in alcohol, drug, and mental disorder comorbidity. Drug Alcohol Depend 1995, 39:129–138.
49. Arata CM, Stafford J, Tims MS: High school drinking and its consequences. Adolescence 2003, 38:567–579.
50. Bellis MA, Hughes K, Morleo M, Tocque K, Hughes S, Allen , Harrison D, Fe-Rodriguez E: Predictors of risky alcohol consumption in schoolchildren and their implications for preventing alcohol-related harm. Subst Abuse Treat Prev Policy 2007, 2:15–26.
51. Sher KJ: Towards a cognitive theory of substance use dependence. In Handbook of Implicit Cognition and Addiction. Edited by: Wiers R, Stacy A. Thousand Oaks, CA: Sage Publications; 2006:273–276.
52. Chard KM: An evaluation of cognitive processing therapy for the treatment of posttraumatic stress disorder related to childhood sexual abuse. J Consult Clin Psychol 2005, 73:965–971.
53. Cloitre M, Koenen KC, Cohen LR, Han H: Skills training in affective and interpersonal regulation followed by exposure: a phase-based treatment for PTSD related to childhood abuse. J Consult Clin Psychol 2002, 70:1067–1074.

54. Lundqvist G, Svedin CG, Hansson K, Broman I: Group therapy for women sexually abused as children: Mental health before and after group therapy. J Interpers Violence 2006, 21:1665–1677.
55. Naimi T, Brewer B, Mokdad A, Denny C, Serdula M, Markes J: Definitions of binge drinking. JAMA 2003, 289:1636.
56. Babor TF, Steinberg K, Anton R, Del Boca F: Talk is cheap: Measuring drinking outcomes in clinical trials. J Stud Alcohol 2000, 61:55–63.
57. Dube SR, Williamson DF, Thompson T, Felitti VJ, Anda RF: Assessing the reliability of retrospective reports of adverse childhood experiences among adult HMO members attending a primary care clinic. Child Abuse Negl 2004, 28:729–737.
58. Hardt J, Rutter M: Validity of adult retrospective reports of adverse childhood experiences: Review of the evidence. J Child Psychol Psychiatry 2004, 45:260–273.
59. Williams LM: Recall of childhood trauma: A prospective study of women's memories of child sexual abuse. J Consult Clin Psychol 1994, 62:1167–1176.
60. Boles SM, Joshi V, Grella C, Wellisch J: Childhood sexual abuse patterns, psychosocial correlates, and treatment outcomes among adults in drug abuse treatment. J Child Sex Abus 2005, 14:39–55.

Relation between Depression and Sociodemographic Factors

Noori Akhtar-Danesh and Janet Landeen

ABSTRACT

Background

Depression is one of the most common mental disorders in Western countries and is related to increased morbidity and mortality from medical conditions and decreased quality of life. The sociodemographic factors of age, gender, marital status, education, immigrant status, and income have consistently been identified as important factors in explaining the variability in depression prevalence rates. This study evaluates the relationship between depression and these sociodemographic factors in the province of Ontario in Canada using the Canadian Community Health Survey, Cycle 1.2 (CCHS-1.2) dataset.

Methods

The CCHS-1.2 survey classified depression into lifetime depression and 12-month depression. The data were collected based on unequal sampling probabilities to ensure adequate representation of young persons (15 to 24) and seniors (65 and over). The sampling weights were used to estimate the prevalence of depression in each subgroup of the population. The multiple logistic regression technique was used to estimate the odds ratio of depression for each sociodemographic factor.

Results

The odds ratio of depression for men compared with women is about 0.60. The lowest and highest rates of depression are seen among people living with their married partners and divorced individuals, respectively. Prevalence of depression among people who live with common-law partners is similar to rates of depression among separated and divorced individuals. The lowest and highest rates of depression based on the level of education is seen among individuals with less than secondary school and those with "other post-secondary" education, respectively. Prevalence of 12-month and lifetime depression among individuals who were born in Canada is higher compared to Canadian residents who immigrated to Canada irrespective of gender. There is an inverse relation between income and the prevalence of depression ($p < 0.0001$).

Conclusion

The patterns uncovered in this dataset are consistent with previously reported prevalence rates for Canada and other Western countries. The negative relation between age and depression after adjusting for some sociodemographic factors is consistent with some previous findings and contrasts with some older findings that the relation between age and depression is U-shaped. The rate of depression among individuals living common-law is similar to that of separated and divorced individuals, not married individuals, with whom they are most often grouped in other studies.

Background

Depression is a significant public health concern worldwide and has been ranked as one of the illnesses having the greatest burden for individuals, families, and society [1,2]. In Canada depression accounts for $14.4 billion annually of health care spending, lost productivity, and premature death [3]. As well, depression is related to increased morbidity and mortality from medical conditions [4-6] and decreased quality of life [7,8] among many other negative consequences.

Given the significant impact of depression on individuals and society as a whole, a comprehensive analysis of the prevalence of depression is necessary to ensure that previous findings remain applicable in today's society. The most recent national data comes from the Canadian Community Health Survey: Mental Health and Well-being (CCHS 1.2) which has found a lifetime prevalence for major depression in 12.2% and major depression occurring in the past 12 months in 4.8% of the population [9]. This is consistent with earlier epidemiological studies of depression in Canada which have found one-year prevalence rates ranging from 4 to 12% [9-11] similar to findings from the United States [12,13]. Demographics of Canada, and particularly of Ontario, are changing with the aging population and the increasing numbers of new immigrants. Recent statistics shows that Ontario receives more than 50 percent of immigrants to Canada [14]. Based on regional changes in the demographic characteristics of the population, it is important to examine regional subsets of the Canadian data on depression.

The sociodemographic factors of age, gender, marital status, education, and income have consistently been identified as important factors in explaining the variability in the prevalence of depression. Key North American studies, particularly the Epidemiologic Catchment Area Study [13], the National Comorbidity Survey [15], the Canadian National Population Health Survey [16], and the Ontario Health Survey [10] found prevalence rates varying from 2.8% to 10.3%, based on the demographic factors of age and gender. Patten and colleagues [9] report that the CCHS 1.2 Canadian-wide survey found significant interactions among age, sex, and marital status, with single women reporting lower rates of depression with increased age and single men reporting increasing rates [9]. It is important for local planners to have a more detailed analysis of the picture of depression in Ontario.

Previous research has found that age is one of the demographic characteristics that accounts for much of the variance in the prevalence of depression. A Canadian National Population Health Survey found that the prevalence of 12-month depression varied in men from "too low to report" for men over 65 to a high of 5.2% for the 12 to 24 age group [16]. Women's prevalence also varied by age, ranging from a low of 3.1% for women over 65 to a high of 9.6% for the 12 to 24 age group [16]. The Ontario Health Survey found comparable variation based on age [10]. This pattern is consistent with findings from Australia [17].

Prevalence for depression has also been found to vary considerably based on gender [18]. Consistently, women have nearly double to triple the prevalence rates for 12-month depression compared to men [10,15-17,19]. There are also gender differences in both the use of outpatient treatment [12] and response to antidepressants [20].

Marital status has been found to interact with gender in accounting for variance in the prevalence of depression. In Australia, those who were separated or divorced had a high rate of anxiety disorders (18%) and affective disorders (12%)

[17]. In Canada, single mothers have been found to have prevalence of 15.4% compared to 6.8% for married mothers [21], although this increase in rate of depression may relate to the demands of parenting rather than on marital status, per se.

Traditional wisdom has long held that there is an association between depression and socioeconomic status (SES). Several recent studies confirm a strong inverse relationship between SES and mental disorder [18,22-24]. Published research indicates that despite differences in definitions and measurements of SES, the likelihood of depression in the lowest SES group is as much as twice that found in the highest SES group [24,25]. People in the lowest class are far more likely to suffer from psychiatric distress than those in the highest class [26]. Lennon et al. [23] concluded that one out of every five women on welfare met standard criteria for major depression. Epidemiological studies of depression in Canada and United States found differences in the prevalence rates of depression based on SES factors [10-13,27]. However, a review by Kohn et al. [28] found that patterns of relationships were not always consistent. Therefore, it is prudent to periodically reassess the relationship between depression and sociodemographic factors because of the changing demographic composition of Ontario.

While effective services and treatments for depression have been identified, the stigma associated with depression has been identified as a barrier to seeking treatment [29]. Worldwide stigma and discrimination have been recognized as major contributors to increasing the burden of mental illness and negative attitudes towards mental illness have been distressingly pervasive [30]. It is known that attitudes toward mental illness (including stigma and discrimination) vary across cultures, and symptoms may also vary while the underlying illness remains the same [30,31]. The cultural mosaic of Canada, and particularly Ontario, is changing as all population growth within the region is now attributable to immigration as the birth rate declines [32]. Thus, it is important to begin to explore the rate of depression in the immigrant population of Ontario. Examining the relationships among the prevalence of depression, immigration status, and demographic factors for the changing population in Ontario is a relevant first step in unraveling some the complex interactions for this significant problem.

In this article we use the dataset from the 2002 Canadian Community Health Survey, Cycle 1.2 (CCHS-1.2) to estimate the prevalence rate of depression in Ontario and whether or not there are differences in subgroups of the population based on the sociodemographic factors of age, gender, marital status, immigration status, education, and income level.

Methods

The Canadian Community Health Survey classified depression into lifetime depression and 12-month depression. One strength of the CCHS 1.2 survey is that it used the Composite International Diagnostic Interview (CIDI) developed by the World Mental Health Project to measure depression [33] with major refinements from the original CIDI, although limited validation studies have been published on the tool [9]. The CIDI was designed to capture cross cultural incidence of mental illness. However, there have been some concerns about the potential for misunderstanding of key concepts used in the survey which might result in an under-reporting from individuals with low education [33]. Every attempt was made to minimize language bias, with interviewers recruited "with a wide range of language competencies. To help these interviewers, an 'official' translation of key terms was created in Chinese and Punjabi, the two most prevalent non-official languages from CCHS Cycle 1.1. Interviewers were restricted from conducting interviews in any other language because of the complexity of the question concepts. Cultural biases toward mental illness could also have led to an under-reporting of depression among immigrant groups.

Respondents who experienced the following criteria associated with major depressive episode (MDE) were classified as being affected by lifetime depression: 1) a period of two weeks or more with depressed mood or loss of interest or pleasure and at least five additional symptoms from the following nine: depressed mood, diminished interest in hobbies or activities, significant weight loss/gain or change in appetite, insomnia or hypersomnia, psychomotor agitation or retardation, fatigue or loss of energy, feelings of worthlessness, diminished ability to think or concentrate, and recurrent thoughts of death, 2) clinically significant distress or social or occupational impairment; and, 3) the symptoms are not better accounted for by bereavement.

Also, respondents who experienced the following criteria associated with MDE were classified as having 12-month depression, 1) meet the criteria for lifetime diagnosis of MDE, 2) report a 12-month episode, and 3) report marked impairment in occupational or social functioning. These definitions are consistent with the classifications of major depression found in the DSM-IV [34]. The standard algorithm for establishing the existence of depression on the CIDI was used and no further restrictions were used except for those indicated above. The additional requirement of meeting clinical significance was not noted in the CCHS 1.2, which has been suggested to minimize any potential over-reporting of mental disorders using the CIDI [33,35].

Data Source

The CCHS-1.2 dataset which includes 12376 respondents in Ontario is based on unequal sampling probabilities due to the design of the study to ensure adequate representation of the sample. One person aged 15 and over was randomly selected from each sampled household. Individuals living in health care institutions, in the military, or living on Indian Reserves were excluded from the survey.

Statistical Analysis

To control for the non-proportional sampling effect of the CCHS-1.2 dataset the proper sampling weights provided by Statistics Canada were used to calculate the percentages of participants in each subgroup of the population and to estimate the prevalence rates of depression. Then, for each prevalence rate, a 95 percent confidence interval (95%CI) is provided using the bootstrap re-sampling program provided by Statistics Canada. Also, the bootstrap program was used in a multiple logistic regression technique to estimate the odds ratio of depression for each demographic and socio-economic factor. The sampling weights were used in conducting chi-square tests and chi-square test for trend. The statistical program SPSS version 15 was used for statistical analysis.

Results

Among the 12376 participants there were 5660 males and 6716 females. Table 1 represents the number and percentages of participants based on the sociodemographic factors. About 64% of the participants were 25–64 years old and 57.1% of them were married and living with their spouses at the time of participation. About 32 percent of the participants immigrated to Canada compared to 68% who were born in Canada. Most of the participants had some post-secondary education (56.5%) and about 40.0% of the participants lived with annual household income of less than $50,000.

In total, the prevalence rates of lifetime depression and 12-month depression are 11.0% (95%CI, 10.2 to 11.7) and 4.8% (95%CI, 4.3 to 5.3) in the province of Ontario. Table 2 presents the prevalence rate of lifetime and 12-month depression based on the sociodemographic factors.

Table 1. Distribution of the sociodemographic variables in the sample

	Male		Female		Total	
	Number	Percent	Number	Percent	Number	Percent
Age group (Year)						
15–19	508	10.0	490	8.5	998	9.2
20–24	391	7.0	500	7.2	891	7.1
25–44	2089	40.0	2345	38.5	4434	39.0
45–64	1692	29.7	1780	29.5	3472	29.6
65–74	570	8.4	848	9.6	1418	9.0
≥75	410	5.0	753	6.8	1163	5.9
Total	**5660**	**100.0**	**6716**	**100.0**	**12376**	**100.0**
Marital status						
Married	2835	58.0	2984	56.2	5819	57.1
Common-law	352	6.7	356	5.1	708	5.9
Widowed	251	2.3	994	8.1	1245	5.2
Separated	219	2.2	300	3.0	519	2.6
Divorced	332	3.2	516	5.1	848	4.2
Single	1665	27.6	1558	22.6	3223	25.0
Total	**5654**	**100.0**	**6708**	**100.0**	**12362§**	**100.0**
Education						
Less than secondary	1417	23.5	1797	24.1	3214	23.8
Secondary graduation	1029	18.0	1309	21.4	2334	19.7
Other post-secondary	491	8.9	586	8.6	1077	8.8
Post-sec graduation	2688	49.6	2990	45.9	5678	47.7
Total	**5625**	**100.0**	**6682**	**100.0**	**12307§**	**100.0**
Immigrant						
Yes	1297	32.2	1601	31.6	2898	31.9
No	4339	67.8	5078	68.4	9417	68.1
Total	5636	100	6679	100	12315	100
Household income						
< $10,000	162	2.0	235	2.3	397	2.2
$10,000–$14,999	214	2.4	527	4.7	741	3.6
$15,000–$19,999	192	2.8	392	4.2	584	3.5
$20,000–$29,999	518	7.5	793	9.7	1311	8.6
$30,000–$39,999	593	10.1	758	11.3	1351	10.7
$40,000–$49,999	562	9.5	608	10.2	1170	9.8
$50,000–$59,999	536	10.1	576	10.8	1112	10.5
$60,000–$79,999	955	18.8	922	17.4	1877	18.1
$80,000 or more	1554	36.7	1302	29.3	2856	33.0
Total	**5286**	**100.0**	**6113**	**100.0**	**11399§**	**100.0**

§ The total number is different from 12376 because of missing or non-applicable data

Depression and Age

The highest prevalence rate of lifetime depression (14.3%) is seen in the age group of 20 to 24 years and the lowest rate (4.3%) in the age group of 75 years and over. Similarly, the highest and lowest rates of 12-month depression are 9.6% and 1.7%, which are seen in the same age groups (Table 2). The prevalence of both types of depression increases with age to the highest level for 20 to 24 year olds and then decreases steadily to its lowest level for the participants aged 75 years and over.

Table 2. Prevalence of lifetime and 12-month depression based on the sociodemographic factors

	Lifetime depression					12-month depression				
	Gender		Total			Gender		Total		
Demographic factor	Male	Female	(%)	95%CI	χ^2 (P-value)	Male	Female	(%)	95%CI	χ^2 (P-value)
Age										
15–19	3.1	12.1	7.4	(5.1, 9.6)	138.980 (<0.001)	1.9	9.6	5.5	(3.5, 7.5)	104.879 (<.001)
20–24	11.1	17.3	14.3	(11.3, 17.3)		7.5	11.5	9.6	(7.0, 12.2)	
25–44	9.1	15.1	12.1	(10.9, 13.3)		3.6	6.6	5.1	(4.3, 5.9)	
45–64	9.8	15.3	12.6	(11.0, 14.2)		3.7	5.4	4.6	(3.5, 5.7)	
65–74	4.5	7.8	6.3	(4.8, 7.7)		1.6	2.1	1.8	(1.1, 2.6)	
≥75	3.6	4.8	4.3	(2.9, 5.7)		1.6	1.7	1.7	(0.8, 2.5)	
Marital status										
Now married	6.0	11.0	8.5	(7.6, 9.5)	301.652	2.0	3.5	2.8	(2.3, 3.3)	
Common-law	18.4	22.3	20.2	(15.4, 25.0)		9.4	10.8	10.0	(5.7, 14.2)	
Widowed	8.8	10.6	10.2	(8.0, 12.0)	(<.001)	4.7	4.3	4.4	(2.8, 6.0)	
Separated	22.6	25.2	24.2	(19.2, 29.2)		10.4	15.4	13.4	(9.4, 17.4)	243.295
Divorced	22.7	27.7	26.1	(21.9, 30.2)		7.6	14.1	11.7	(8.6, 14.7)	(<.001)
Single	7.3	14.7	10.7	(9.3, 12.2)		3.9	9.2	6.3	(5.1, 7.6)	
Immigrant										
Yes	6.5	9.1	7.8	(6.3, 9.4)	60.699	3.2	4.8	4.0	(3.0, 5.0)	**19.619**
No	9.1	15.8	12.4	(11.5, 13.3)	(<.001)	3.6	6.6	5.4	(4.5, 5.7)	(<.001)
Education										
< Secondary	7.1	11.1	9.1	(7.7, 10.6)	23.241 (<0.001)	3.7	6.3	5.0	(3.9, 6.1)	9.586 (=.022)
Secondary grad.	8.8	12.1	10.6	(8.9, 12.3)		3.0	5.2	4.2	(3.2, 5.2)	
Other post-sec.	11.1	15.6	13.4	(10.5, 16.2)		6.1	6.5	6.3	(4.4, 8.2)	
Post-sec. grad.	8.0	15.2	11.5	(10.4, 12.6)		3.1	6.3	4.7	(3.9, 5.5)	
Household income										
<$10,000	16.9	19.7	18.4	(13.4, 23.4)	χ^2_{Trend} = 51.24 (<.0001)	9.3	13.1	11.3	(7.3, 15.3)	χ^2_{Trend} = 103.08 (<.0001)
$10,000–$14,999	11.2	18.5	16.1	(13.0, 19.1)		6.2	10.2	8.9	(6.5, 11.3)	
$15,000–$19,999	13.0	17.5	15.7	(11.4, 19.9)		8.5	9.2	8.9	(5.9, 12.0)	
$20,000–$29,999	8.4	13.5	11.3	(9.2, 13.4)		4.5	7.2	6.0	(4.4, 7.7)	
$30,000–$39,999	8.0	11.9	10.1	(8.0, 12.2)		2.8	5.5	4.3	(2.9, 5.6)	
$40,000–$49,999	9.0	12.1	10.6	(8.4, 12.8)		3.3	4.1	3.7	(2.4, 5.0)	
$50,000–$59,999	7.5	13.4	10.5	(8.2, 12.9)		3.4	4.2	3.8	(2.3, 5.3)	
$60,000–$79,999	7.0	12.7	9.7	(8.2, 11.3)		2.8	5.1	3.9	(2.9, 5.0)	
≥$80,000	7.9	14.0	10.6	(9.1, 12.2)		2.8	4.9	3.8	(2.7, 4.9)	
Total	**8.2**	**13.7**	**11.0**	**(10.2, 11.7)**		**3.5**	**6.1**	**4.8**	**(4.3, 5.3)**	

Depression and Gender

Women suffer more from both types of depression than men. The highest and lowest rates of lifetime and 12-month depression for both men and women were recorded in the age groups of 20 to 24 years and 75 and over, respectively (Table 2).

Depression and Marital Status

The prevalence of depression varies with the marital status. The highest rates of lifetime and 12-month depression are seen in divorced and separated respondents, respectively. The lowest rate for both types of depression is seen among married people. Also, a high rate is seen among individuals living with "common-law" partners where the term "common-law" refers to the living of a man and a woman together in a marital status without legal action (Table 2).

Depression and Immigrant Status

In general prevalence of depression among individuals who were born in Canada is higher compared to Canadian residents who immigrated to Canada irrespective of gender and type of depression (Table 2).

Depression and Education

Respondents whose education level was less than secondary school have the lowest rate of lifetime depression (9.1%); and the highest rate of lifetime depression (13.4%) is seen among those with "other post-secondary" education. A similar pattern is seen for 12-month depression. Although these results indicate that the prevalence of depression differs based on the level of education, there is no linear pattern for this relationship. For both lifetime and 12-month depression the prevalence rate was higher for "other post secondary education" than "post-secondary education."

Depression and Household Income Level

The highest prevalence rate of lifetime depression (18.4%) is seen in households with an income level of less than $10,000 per year. The prevalence of lifetime depression then decreases as the income increases. The same pattern is observed for 12-month depression with the highest rate of 11.3% in households with the income of less than $10,000 per year. However, there seems to be an threshold effect as the prevalence rate decreases much faster for income level of up to $30,000 than for $30,000 and over which will be further elaborated in the modeling section. The chi-square test for trend indicates that for both types of depression there is a noticeable inverse relationship between the level of income and the prevalence of depression ($p < 0.0001$; Table 2).

Table 3. Relationship between depression and the sociodemographic factors

Variable	B	SE (B)	P-value	OR‡	95% CI for OR	
					Lower	Upper
A: Lifetime depression						
Age/10§	-0.10	0.03	<0.0001	0.90	0.86	0.95
Gender						
Female†				1.00		
Male	-0.53	0.09	<0.0001	0.59	0.50	0.70
Marital Status						
Single†				1.00		
Married	0.00	0.12	0.998	1.00	0.78	1.28
Common-law	0.90	0.18	<0.0001	2.46	1.73	3.51
Widowed	0.26	0.19	0.182	1.30	0.89	1.90
Separated	0.97	0.17	<0.0001	2.65	1.91	3.67
Divorced	1.20	0.15	<0.0001	3.33	2.46	4.50
Education						
Less than secondary†				1.00		
Secondary	0.23	0.14	0.096	1.26	0.96	1.66
Post-secondary	0.43	0.12	0.0003	1.54	1.22	1.93
Household income	-0.20	0.14	<0.0001	0.82	0.74	0.90
Immigrant status (yes, no)	-0.52	0.14	0.0001	0.60	0.46	0.78
Constant	-1.05	0.23	<0.0001	0.35	0.22	0.55
B: 12-month depression						
Age/10§	-0.19	0.05	<0.0001	0.83	0.76	0.90
Gender						
Female†				1.00		
Male	-0.46	0.14	0.0008	0.63	0.48	0.83
Marital Status						
Single†				1.00		
Married	-0.36	0.19	0.0651	0.70	0.48	1.02
Common-law	0.89	0.28	0.0016	2.43	1.40	4.23
Widowed	0.18	0.32	0.5713	1.20	0.64	2.24
Separated	1.16	0.23	<0.0001	3.20	2.05	4.97
Divorced	1.01	0.23	<0.0001	2.76	1.75	4.33
Education						
Less than secondary†				1.00		
Secondary	0.06	0.19	0.753	1.06	0.73	1.55
Post-secondary	0.30	0.17	0.073	1.35	0.97	1.86
Household income	-0.35	0.59	<0.0001	0.71	0.63	0.79
Immigrant status (yes, no)	-0.19	0.19	0.3266	0.83	0.57	1.20
Constant	-0.99	0.32	0.0022	-	0.19	0.70

‡ Odds ratio, § Age divided by 10, † Reference group

Modeling Depression based on the Sociodemographic Factors

We performed a logistic regression analysis for lifetime and 12-month depression to identify the most important sociodemographic factors associated with depression. Gender and marital status were considered as categorical variables with "female" and "single" as reference groups for gender and marital status, respectively. Education was dealt with as a categorical variable while "other post secondary education" and "post-secondary education" were combined into one group and the "less than secondary education" group was considered as the reference group. The immigrant status was categorized as 'yes' and 'no' identifying who immigrated to Canada and those who were born in Canada (the reference group). Table 2 indicates a strong inverse relation between household income and depression up to income level of $30,000. Then, the prevalence rate becomes approximately stable. Based on this finding we classified the income level of $30,000 and more into one category for the modeling purposes. The following noticeable results emerged from this analysis (Table 3):

1. The odds of being affected by lifetime and 12-month depression for men is about 0.60 times of that for women.
2. Marital status emerged as an important predictor for both types of depression:
   a. The odds of being affected by lifetime depression for married persons is about the same as that for single individuals. For 12-month depression the odds ratio is 0.70, although it is not significantly different from one (95%CI, 0.48 to 1.02).
   b. The odds of lifetime or 12-month depression among individuals who live with common-law partners is about 2.5 times of that for singles.
   c. The odds of being affected by lifetime depression for separated and divorced individuals is more than 2.5 times of that for singles; odds ratio is 2.65 (95%CI, 1.91 to 3.67) for separated and 3.33 (95%CI, 2.46 to 4.50) for divorced individuals. For 12-month depression these odds ratios are 3.20 (95%CI, 2.05 to 4.97) and 2.76 (95%CI, 1.75 to 4.33), respectively.
3. The odds of living with lifetime depression among individuals with any kind of post-secondary education is 1.54 times compared to individuals with less than secondary education (95%CI: 1.22 to 1.93). For 12-month depression this odds is not statistically significant (odds ratio = 1.35; 95%CI: 0.97 to 1.86).
4. Income has a significant association with both lifetime and 12-month depression. For the income level of up to $30,000 the odds ratio of lifetime depression for each $10000 increase in income is 0.82 (95%CI, 0.74 to 0.90). Similarly, for 12-month depression the odds ratio is 0.71 (95%CI, 0.63 to 0.79).
5. Prevalence rate of lifetime depression among immigrants is 60% of that among Canadian born individuals. For the 12-month depression the odds ratio is 0.83 but it is not statistically significant (95%CI, 0.57 to 1.20).

Discussion

The findings of this analysis illustrate that despite changes in the overall demographics of the Ontario population, the prevalence rates of depression remain consistent with previous Canadian and American epidemiological surveys [10,15]. The patterns uncovered in this analysis are at the higher limits of the previously reported prevalence rates for Canada, and in particular for the province of Ontario [10]. The rates for depression in Ontario are consistent with the results from the larger Canadian sample from which this dataset has been extracted.[9]. This finding was anticipated given that Ontario represents about 40% of the total Canadian population [36]. This study has further described current distribution of depression based on age, gender, marital status, education, immigrant- nonimmigrant status, and household income.

Overall, our findings are consistent with some previous findings that women have double to triple the prevalence rates for 12-month depression compared to men [10,15,17,19], however, the age-specific rates in Table 2 indicate that gender differences diminish with age and there is virtually no difference at age 75+ which is in agreement with Gutiérrez-Lobos et al. [37].

We observed a negative relation between age and depression for both lifetime and 12-month depression after adjusting for some sociodemographic factors using logistic regression technique. These results are consistent with some previous findings [38-40] which rule out some older findings that relation between age and depression is U-shaped with the lowest reported levels of depression at ages 45–49 [41,42].

The relation between depression and marital status is highly significant. While our analysis confirms previous reported patterns for depression based on marital status [43], one notable difference emerged. The prevalence of depression in individuals living common-law was similar to that of separated and divorced individuals, not married individuals with whom they are most often grouped in other studies, including the recent report using the same dataset for all Canada [9]. This finding challenges the common practice of combining married and common-law individuals in the same category and suggests that these groups should be analyzed separately until the consistency of this finding is upheld or refuted.

While rates of depression in individuals born in Canada were higher than for immigrants, depression is still a health concern for the immigrant group. We found that immigrant status was highly related to lifetime depression but not to depression in the past year. Because this was a cross-sectional survey it is impossible to know if the depression occurred before or after immigration. Furthermore, the only data available on immigrant status was whether or not the individual was born in Canada. While attempts were made to minimize language barriers for data collection, it would be very helpful to know whether the respondents were refuges or skilled immigrants and how recently they immigrated. This survey only provides the starting point for exploring depression in immigrants in Ontario, and future studies are required to gain a better understanding of the complexities of factors that may contribute to depression in this group. Furthermore, given the cross-cultural variation in attitudes toward depression [44], the findings may not reflect accurate depression rates amongst recent immigrants.

Socioeconomic status, as indicated by education and income, also showed significant association with depression. In a multivariate analysis using logistic regression analysis, these variables showed strong relationships with depression.

Nevertheless, it remains important to target programs for those at the lowest income level, particularly to women between the ages of 20 and 64 and people who are divorced or separated. In the light of this analysis, individuals who live with "common-law" partners need special attention.

In conclusion, this study has provided a new snapshot of the prevalence of depression in Ontario. While this is not very different from what has been found before, we have provided details of subgroups of the population who are most at risk for depression. These findings are potentially important as they are from a large random sample of respondents. The results show significant relationships between depression and different sociodemographic factors. The results confirm the findings from other Canadian and international studies [10,11] and add strong weight toward the confirmation of such relations.

One limitation of the CCHS-1.2 dataset might be that it provides self-reported information from a cross-sectional study. Furthermore, because of funding and space limitations, we analyzed only Ontario data, and replicating this analysis in all of Canada would be beneficial to health policy makers.

Competing Interests

The author(s) declare that they have no competing interests.

Authors' Contributions

Both authors contributed equally in designing the study and drafting the manuscript. The analysis was carried out by NAD. Both authors read and approved the final manuscript.

Acknowledgements

We are grateful to Alina Dragan for helping in analysis of the dataset. This research was supported by the grant number 04-059 from the Ministry of Health and Long-Term Care of Ontario, Canada. While the research and analysis are based on data from Statistics Canada, the opinions expressed do not represent the views of Statistics Canada.

References

1. U.S. Department of Health and Human Services: Mental health: A report of the surgeon general. Rockville, MD; 1999.
2. World Health Organization: Prevention and promotion in mental health. Geneva. 2002.
3. Health Canada: A proactive approach to good health. 2001.
4. Sartorius N, Ustun TB, Lecrubier Y, Wittchen HU: Depression comorbid with anxiety: Results from the WHO study on Psychological Disorders in Primary Health Care. British Journal of Psychiatry 1996, 168:38–43.
5. Nuyen J, Volkers AC, Verhaak PF, Schellevis FG, Groenewegen PP, Van den Bos GA: Accuracy of diagnosing depression in primary care: the impact of chronic somatic and psychiatric co-morbidity. Psychol Med 2005, 35:1185–1195.
6. Ramasubbu R, Patten SB: Effect of depression on stroke morbidity and mortality. Can J Psychiatry 2003, 48:250–257.
7. D'Alisa S, Miscio G, Baudo S, Simone A, Tesio L, Mauro A: Depression is the main determinant of quality of life in multiple sclerosis: a classification-regression (CART) study. Disabil Rehabil 2006, 28:307–314.
8. Fruhwald S, Loffler H, Eher R, Saletu B, Baumhackl U: Relationship between depression, anxiety and quality of life: a study of stroke patients compared to chronic low back pain and myocardial ischemia patients. Psychopathology 2001, 34:50–56.
9. Patten SB, Wang JL, Williams JV, Currie S, Beck CA, Maxwell CJ, El Guebaly N: Descriptive epidemiology of major depression in Canada. Can J Psychiatry 2006, 51:84–90.
10. Offord DR, Boyle MH, Campbell D, Goering P, Lin E, Wong M, Racine YA: One-year prevalence of psychiatric disorder in Ontarians 15 to 64 years of age. Canadian Journal of Psychiatry 1996, 41:559–563.

11. Patten SB: Progress against major depression in Canada. Canadian Journal of Psychiatry 2002, 47:775–780.

12. Olfson M, Marcus SC, Druss B, Elinson L, Tanielian T, Pincus HA: National trends in the outpatient treatment of depression. Journal of the American Medical Association 2002, 287:203–209.

13. Robins LN, Regier DA: Psychiatric disorders in America: the Epidemiologic Catchment Area Study. New York: The Free Press; 1991.

14. Ontario Ministry of Finance: Ontario Demographic Quarterly. 2006.

15. Kessler RC, Mcgonagle KA, Zhao SY, Nelson CB, Hughes M, Eshleman S, Wittchen HU, Kendler KS: Lifetime and 12-Month Prevalence of DSM-III-R psychiatric-disorders in the United-States–Results from the National-Comorbidity-Survey. Archives of General Psychiatry 1994, 51:8–19.

16. Patten SB: Incidence of major depression in Canada. Canadian Medical Association Journal 2000, 163:714–715.

17. Australian Bureau of Statistics [http:/ / www.abs.gov.au/ Ausstats/ abs@.nsf/ Lookup/ 670D4F8706B05404CA2568A900136280], 2006.

18. Wade TJ, Cairney J, Pevalin DJ: Emergence of gender differences in depression during adolescence: national panel results from three countries. J Am Acad Child Adolesc Psychiatry 2002, 41:190–198.

19. Kornstein SG, Schatzberg AF, Thase ME, Yonkers KA, McCullough JP, Keitner GI, Gelenberg AJ, Ryan CE, Hess AL, Harrison W, et al.: Gender differences in chronic major and double depression. Journal of Affective Disorders 2000, 60:1–11.

20. Kornstein SG, Schatzberg AF, Thase ME, Yonkers KA, McCullough JP, Keitner GI, Gelenberg AJ, Davis SM, Harrison WM, Keller MB: Gender differences in treatment response to sertraline versus imipramine in chronic depression. American Journal of Psychiatry 2000, 157:1445–1452.

21. Cairney J, Thorpe C, Rietschlin J, Avison WR: 12-month prevalence of depression among single and married mothers in the 1994 National Population Health Survey. Canadian Journal of Public Health-Revue Canadienne de Sante Publique 1999, 90:320–324.

22. Eaton WW, Muntaner C, Bovasso G, Smith C: Socioeconomic status and depressive syndrome: the role of inter- and intra-generational mobility, government assistance, and work environment. J Health Soc Behav 2001, 42:277–294.

23. Lennon MC, Blome J, English K: Depression among women on welfare: A review of the literature. Depression among women on welfare: A review of the literature 2002, 57:27–32.

24. Lorant V, Deliege D, Eaton W, Robert A, Philippot P, Ansseau M: Socioeconomic inequalities in depression: A meta-analysis. American Journal of Epidemiology 2003, 157:98–112.
25. Fortenberry JD: Socioeconomic status, schools, and adolescent depression: Progress in the social epidemiology of adolescent health. Journal of Pediatrics 2003, 143:427–429.
26. Goode E: Deviant Behavior. Upper Saddle River: Prentice Hall; 1997.
27. Murphy JM, Olivier DC, Monson RR, Sobol AM, Federman EB, Leighton AH: Depression and anxiety in relation to social status. A prospective epidemiologic study. Arch Gen Psychiatry 1991, 48:223–229.
28. Kohn R, Dohrenwend B, Mirotznik H: Edidemilogical findings on selected psychaitric disordrs in the general population. In Adversity, stress and psychopathology. Edited by: Dohrenwend B. New York: Oxford University Press; 1998:235–284.
29. Halter MJ: The stigma of seeking care and depression. Archives of Psychiatric Nursing 2004, 18:178–184.
30. World Health Organization: The World health report: Mental health: new understanding, new hope. Geneva. 2001.
31. Karasz A: Cultural differences in conceptual models of depression. Soc Sci Med 2005, 60:1625–1635.
32. Statistics Canada: The Daily: Canada's population as of July 1, 2006. Ottawa. 2006.
33. Kessler RC, Ustun TB: The World Mental Health (WMH) Survey Initiative Version of the World Health Organization (WHO) Composite International Diagnostic Interview (CIDI). International Journal of Methods in Psychiatric Research 2004, 13:93–121.
34. American Psychiatric Association: Diagnostic and statistical manual of mental disorders. Washington, DC: American Psychiatric Association; 2000.
35. Kessler RC, Abelson J, Demler O, Escobar JI, Gibbon M, Guyer ME, Howes MJ, Jin R, Vega WA, Walters EE, et al.: Clinical calibration of DSM-IV diagnoses in the World Mental Health (WMH) version of the World Health Organization (WHO) composite international diagnostic interview (WMH-CIDI). International Journal of Methods in Psychiatric Research 2004, 13:122–139.
36. Statistics Canada: Canada's population. 2006.
37. Gutierrez-Lobos K, Scherer M, Anderer P, Katschnig H: The influence of age on the female/male ratio of treated incidence rates in depression. BMC Psychiatry 2002, 2:3.

38. Streiner DL, Cairney J, Veldhuizen S: The epidemiology of psychological problems in the elderly. Can J Psychiatry 2006, 51:185–191.

39. Wade TJ, Cairney J: Age and depression in a nationally representative sample of Canadians: a preliminary look at the National Population Health Survey. Can J PublicHealth 1997, 88:297–302.

40. Wade TJ, Cairney J: The effect of sociodemographics, social stressors, health status and psychosocial resources on the age-depression relationship. Can J Public Health 2000, 91:307–312.

41. Kessler RC, Foster C, Webster PS, House JS: The relationship between age and depressive symptoms in two national surveys. Psychol Aging 1992, 7: 119–126.

42. Newmann JP: Aging and depression. Psychol and Aging 1989, 4:150–165.

43. Scarinci IC, Beech BM, Naumann W, Kovach KW, Pugh L, Fapohunda B: Depression, socioeconomic status, age, and marital status in black women: a national study. Ethn Dis 2002, 12:421–428.

44. Karasz A: Cultural differences in conceptual models of depression. Soc Sci Med 2005, 60:1625–1635.

Satisfaction with Life and Opioid Dependence

Jason Luty and Sujaa Mary Rajagopal Arokiadass

ABSTRACT

Background

Serious substance misuse and dependence is widely seen as damaging to an individual and to society in general. Whereas the medical and society effects of substance misuse are widely described, some commentators suggest substance misuse may be an "alternative lifestyle."

Aim

To assess general life satisfaction amongst treatment-seeking people with substance dependence.

Methods

The Satisfaction With Life Scale (SWLS) was administered to a sample of opioid-dependent people receiving substitute medication.

Results

105 subjects and 105 age-sex matched subjects in a comparison group completed the questionnaire. The mean SWLS score was 7.12 (SD = 10.6; median = 6) for patients compared to 22.6 (SD = 6.8) in the comparison group. (Two sided p < 0.0001; Median difference = -13.5; Wilcoxon signed rank test.)

Conclusion

The study used a validated instrument and objective reports to confirm significantly higher rates of dissatisfaction with life among opioid dependent people in treatment when compared to members of the general population.

Introduction

The British Government commissioned several reports in the 1950s in regard to what are often termed "victimless crimes." Recommendations from these reports into prostitution, abortion and homosexuality were gradually introduced (with repeal of the relevant legislation). However the recommendations regarding more liberal legislation on illicit drug use (The Wootton Report, 1969) was never fully introduced [1,2]. Consideration was given at both committees that illicit drug use could be regarded as an alternative lifestyle rather than a criminal deviance or disease. If this were the case, it would be possible that illicit drug users might have the same overall satisfaction with their life as other members of the public. (Clearly, however, social disapproval and legal sanctions might cause lifestyle problems that were unrelated to the acquisition and use of addictive drugs.) One question facing many authorities remains, can illicit drug use be regarded as a valid, although reckless, lifestyle choice (much like rock climbing or motor sport)? To better inform policy decisions on substance misuse a survey was conducted of satisfaction with life of people in treatment for illicit drug problems and members of the public.

Serious substance misuse and dependence is widely seen as damaging to an individual and to society in general [3-8]. Whereas the medical and societal effects of substance misuse are widely described, some commentators suggest substance misuse may be an "alternative lifestyle" [9,10]. Emotional and behavioural problems, including delinquency, truancy and hyperactivity, have repeatedly been found to be associated with and predict substance misuse [11-13].

The aim of the current research was to use a validated instrument to assess general life satisfaction amongst treatment-seeking people with substance (opioid) dependence.

Methods

Participants were approached by researchers at three drug and alcohol services in South East England. Participants were included who were dependent on illicit opioids drugs in receipt of substitute medication. All patients subject to review by medical staff were approached and information sheets were also distributed at the reception of the clinic. The services currently provide substitute prescriptions to approximately 600 patients. The inclusion criteria were: currently receiving treatment for opioid dependence and ability to give written informed consent. There were no specific exclusion criteria. Illicit drug dependence was confirmed using the Minnesota Student Survey Screening [14]. Comparison group subjects were age and sex matched to within 5 years. Comparison group subjects were recruited from the general UK population from a database created as part of another study [15]. They were recruited by direct mail shots sent to addresses at random and newspaper advertisements throughout the UK. The project was approved by the local research ethics committee. All participants completed the Satisfaction With Life Scale [16]. However only patients had urine drug screens performed. Patients also completed some questions on their involvement with the police, school engagement and clinical information such as the duration of illicit drug use. This information was not obtained for the non-drug using comparison group.

The Satisfaction With Life Scale (SWLS; [16,17]) is an extensively validated 6-item self-completion instrument (score 1 to 7). Responses are scored on a 6-point Likert-type scale yielding a maximum overall score range from 5–35.

Results

One hundred and five subjects completed the questionnaire. These were paired with 105 age and sex-matched comparison subjects drawn from the general UK population. Comparison subjects were matched by age to within 5 years. The 105 patients included in the study had a mean age of 33.3 (SD = 12.8) years; 80% were male; 10% were in paid employment; 96% described themselves as white British. Comparison subjects had a mean age of 34.8 (SD = 18.6) years; 76% were in paid employment. In practice, the majority of the patients were receiving long-term prescriptions (in excess of 2 years) although many were attempting to gradually detoxify. 45% were in receipt of buprenorphine prescriptions, the remainder received methadone. The mean age at first heroin use was 21.3 (SD = 7) years and they reported using opiates regularly for 14.2 (SD = 9) years.

The mean SWLS score was 7.12 (SD = 10.6; median = 6) for patients compared to 22.6 (SD = 6.8) in comparison subjects. (Two sided $p < 0.0001$; Median difference = -13.5; Wilcoxon signed rank test.)

The SWLS scores for the comparison subjects were not significantly different from those from the original validation study. (The mean score was 23.5 (SD = 6.43) in 176 US undergraduates in the original validation report; [16]).

Forty per cent of patients were expelled from school while 32 (64%) received no formal qualifications–this is loosely comparable to not completing US high school. Patients obtained a mean of 1.88 (SD = 2.62; median = 0; n = 18) GCSEs or O-levels (qualification obtained at the age of 16 in the UK). The mean age of first contact with the police in patients was 13.3 (SD = 3.0 n = 48) years. The mean age of first use any illicit drug use in patients was 16.1 (SD = 4.5) years. The mean age of first use of heroin or cocaine in patients was 19.5 (SD = 5.3) years. Ninety six per cent of patients reported problems with the police in adolescence.

Discussion

The report clearly shows that opioid dependent people who are in treatment have much lower levels of satisfaction than members of the comparison group. Happiness and "satisfaction" with life are global concepts with philosophical and psychological components [16-18]. The results refer to a treatment seeking population–although these are likely to be representative of those people dependent on illicit opiates in the UK (the majority of who are in contact with treatment services [19]), this would not represent those who tend not to access treatment including people who use infrequently or those who use other illicit drugs such as cannabis or stimulants.

There are many potential determinants of satisfaction with life. These include personality, social expectations, socio-economic factors especially relative deprivation, relationships with significant others (neighbours, parents and children), physical and psychological health, accommodation, employment and problem with authority [20,21]). Moreover there is overwhelming evidence of the damaging effects of illicit drug dependence on both the physical and mental health of users and also on their relationships and social functioning [3,9]. Three potential explanations can be cited. Firstly, opiate dependence leads to chronic mental health problems and physical illness that directly cause dis-satisfaction ("dis-ease"). Secondly, opiate dependence causes secondary social and relationship problems that prevent people achieving their desired goals (e.g. criminality restricts employment; substance misuse damages relationships with family and significant others). Thirdly, it remains possible opiate dependent people have behavioural and psychological traits that prevent them achieving "happiness"–that is these people would remain dissatisfied with life regardless of whether they became substance abusers or not. Whereas it is extremely difficult to disentangle these competing theories, the results presented here clearly show that people with substance misuse problems, even those in treatment, are generally dissatisfied with life and "unhappy."

In relation to the second hypothesis (that dissatisfaction results from secondary social problems resulting indirectly from substance use and acquisition), there are very many possible causes including predisposing factors (history of depression, conduct disorder, poverty in the home) and current concomitant factors (lack of employment, poverty, health, legal sanctions). Some of these may be amenable to intervention and many drug treatment services aim to provide these, including the services from which the patients were recruited. These include advice on benefits and employment and assistance with housing as well as treatment for depression. A comprehensive model for holistic assessment and treatment of substance using people is described by UK government guidelines and are enacted, at least in theory, in all NHS facilities [3,19]. Despite these attempts at resolving the many social difficulties, the patients in treatment remained dissatisfied with life. There has been a longstanding debate regarding the potentially damaging effects of rendering addictive drugs illegal and requests for decriminalizing drugs [1,2]. However any actions in this respect are based primarily on political rather than scientific grounds and these are probably unlikely in the current political climate [22].

Strengths and Limitations

The Satisfaction with Life Scale is a well-validated questionnaire that has been compared to several other instruments and has good psychometric properties [16-18].

Opioid-dependent people may be motivated to seek treatment as a result of dissatisfaction with life. It remains possible that a proportion of opiate users are satisfied with life and do not seek treatment. However other research suggests that at around 80% of illicit opioid users have been in contact with treatment services and at least half are in contact at any one time [23,24].

Conclusion

The study used a validated instrument and objective reports to confirm significantly higher rates of dissatisfaction with life among opioid dependent people in treatment when compared to members of the general population.

Competing Interests

The author(s) declare that they have no competing interests.

Authors' Contributions

Both authors were fully and actively involved in all parts of the project including design, data collection, analysis and manuscript preparation. All authors read and approved the final manuscript.

References

1. Home Affairs Select Committee: [http:/ / www.publications.parliament.uk/ pa/ cm200102/ cmselect/ cmhaff/ 318/ 31802.htm]. The Government's Drugs Policy: Is It Working? Third Report. London: The Stationary Office; 2000. Accessed 04/04/2008
2. Blanchard S: How cannabis was criminalized. [http://www.idmu.co.uk/historical.htm]. London: Independent Drug Monitouring Unit; 2007. Accessed 04/01/2008
3. Department of Health: Drug misuse and dependence–guidelines on clinical management. London: The Stationary Office; 1999.
4. Regier DA, Farmer ME, Rae DS, Locke BZ, Keith SJ, Judd LL, Goodwin FK: Comorbidity of mental disorders with alcohol and other drug abuse: results from the Epidemiological Catchment Area (ECA) study. Journal of the American Psychiatric Association 1990, 264:2511–2518.
5. Brooner RK, King VL, Kidorf M, Schmidt CW Jr, Bigelow GE: Psychiatric and substance use comorbidity among treatment-seeking opioid abusers. Archives of General Psychiatry 1997, 54(1):71–80.
6. Farrell M, Howes S, Bebbington P, Brugha T, Jenkins R, Lewis G, Marsden J, Taylor C, Meltzer : Nicotine, alcohol and drug dependence and psychiatric comorbidity. British Journal of Psychiatry 2001, 179:432–437.
7. Jaycox LH, Morral AR, Juvonen J: Mental health and medical problems and service use among adolescent substance users. Am Acad Child Adolescent Psychiatry 2003, 42:701–9.
8. Brown BertramS: Drugs and Public Health: Issues and Answers The ANNALS of the American Academy of Political and Social Science. 1975, 417(1):110–119.
9. Royal College of Psychiatrists: Drug: dilemmas and choices. London: Gaskell; 2000.
10. Webb E, Ashton CH, Kelly P, Kamah F: An update on British medical students' lifestyles, ncbi.nlm.nih.gov. Med Educ 1998, 32(3):325–31.

11. McAra L: Truancy, School Exclusion and Substance Misuse. Centre for Law and Society, University of Edinburgh, law.ed.ac.uk. 2004, 4:3.
12. Poikolainen K: Antecedents of substance use in adolescence. Current Opinion Psychiatry 2002, 15:241–5.
13. Poikolainen K: Ecstasy and the antecedents of illicit drug use, Anxiety and depression may be risk factors for using ecstasy. BMJ 2006, 332:803–804.
14. Fulkerson JA, Harrison PA, Beebe TJ: DSM-IV substance abuse and dependence. Addiction 1999, 94:495–506.
15. Luty J, Fekadu D, Umoh O, Gallagher J: Validation of a short instrument to measure stigmatised attitudes towards mental illness. Psychiatric Bulluletin 2006, 30:257–260.
16. Diener E, Emmons PA, Larsden RJ, Griffin S: "The satisfaction with life scale." Journal of Personality Assessment 1985, 45:71–75.
17. Pavot W, Diener E, Colvin CR, Sandvik E: Further validation of the Satisfaction with Life Scale: evidence for the cross-method convergence of well-being measures. Journal of Personality Assessment, 1991–ncbi.nlm.nih.gov 1991, 57(1):149–61.
18. Gow AJ, Whiteman MC, Pattie A, Whalley L, Starr J, Deary IJ: Lifetime intellectual function and satisfaction with life in old age. British Medical Journal 2006, 331:141–142.
19. National Treatment Agency [http://www.nta.nhs.uk] 2007. Accessed 04/01/2008
20. Fox CR, Kahneman D: Pancultural explanations for life satisfaction: adding relationship harmony to self-esteem. Journal Personality and Social Psychology 1997, 73(5):1038–51.
21. Schimmack U, Radhakrishnan P, Oishi S, Dzokoto V, Ahadi S: Culture, personality, and subjective well-being: integrating process models of life satisfaction. Journal Personality and Social Psychology 2002, 82(4):582–93.
22. Jones B, Kavanagh D, Moran M, Norton P: Politics UK. London: Pearson Longman; 2006.
23. Rounsaville BJ, Kleber HD: Untreated opiate addicts. Arch Gen Psychiatry 1985, 42:1072–1077.
24. Luty JS: Controlled Survey of Social Problems, Psychological Well-Being and Childhood Parenting Experiences in a Community Sample of Heroin Addicts in Central London. Substance Use and Misuse 2002, 38:46–54.

Can Playing the Computer Game "Tetris" Reduce the Build-Up of Flashbacks for Trauma? A Proposal from Cognitive Science

Emily A. Holmes, Ella L. James, Thomas Coode-Bate and Catherine Deeprose

ABSTRACT

Background

Flashbacks are the hallmark symptom of Posttraumatic Stress Disorder (PTSD). Although we have successful treatments for full-blown PTSD, early interventions are lacking. We propose the utility of developing a 'cognitive vaccine' to prevent PTSD flashback development following exposure to trauma. Our theory is based on two key findings: 1) Cognitive science suggests that the brain has selective resources with limited capacity; 2) The neurobiology of memory suggests a 6-hr window to disrupt memory consolidation. The rationale for a 'cognitive vaccine' approach is as follows: Trauma flashbacks are sensory-perceptual, visuospatial mental images. Visuospatial cognitive tasks selectively compete for resources required to generate mental images. Thus, a visuospatial computer game (e.g. "Tetris") will interfere with flashbacks. Visuospatial tasks post-trauma, performed within the time window for memory consolidation, will reduce subsequent flashbacks. We predicted that playing "Tetris" half an hour after viewing trauma would reduce flashback frequency over 1-week.

Methodology/Principal Findings

The Trauma Film paradigm was used as a well-established experimental analog for Post-traumatic Stress. All participants viewed a traumatic film consisting of scenes of real injury and death followed by a 30-min structured break. Participants were then randomly allocated to either a no-task or visuospatial ("Tetris") condition which they undertook for 10-min. Flashbacks were monitored for 1-week. Results indicated that compared to the no-task condition, the "Tetris" condition produced a significant reduction in flashback frequency over 1-week. Convergent results were found on a clinical measure of PTSD symptomatology at 1-week. Recognition memory between groups did not differ significantly.

Conclusions/Significance

Playing "Tetris" after viewing traumatic material reduces unwanted, involuntary memory flashbacks to that traumatic film, leaving deliberate memory recall of the event intact. Pathological aspects of human memory in the aftermath of trauma may be malleable using non-invasive, cognitive interventions. This has implications for a novel avenue of preventative treatment development, much-needed as a crisis intervention for the aftermath of traumatic events.

Introduction

We suggest that basic principles from cognitive science may be used to help develop an intervention for trauma flashbacks, and propose a 'cognitive vaccine' approach. That is, that the delivery of specific cognitive tasks may help 'inoculate' against the escalation of flashbacks after a traumatic event. Post Traumatic Stress Disorder (PTSD) is a psychiatric disorder that can result from experiencing or viewing a traumatic event involving death, serious injury, or threat to self or others [1], [2]. A precursor [3] and indeed the hallmark symptom of PTSD [1] is vivid flashbacks to the trauma, that is, distressing, re-experiencing of the trauma in the form of intrusive, image-based, sensory-perceptual memories. For example, following a motor vehicle accident, a person may later experience intrusive flashbacks where in their mind's eye they suddenly see a vision of a looming car accompanied by the sound of crashing metal.

Although we have successful treatments for full-blown PTSD, crisis interventions to reduce the build up of symptoms in the early aftermath of trauma are lacking. Current interest in the manipulation of memories post-trauma is particularly focused on pharmacological means, for example, propranolol administration [4]. In addition to the potential for side-effects with pharmacological approaches, there are potential ethical concerns if voluntary memories for human experience are suppressed [5]. For example, removing flashbacks at the expense of being able to deliberately remember what happened during a trauma could compromise a trauma victim's ability to testify in court. We have also raised clinical concern over treatment innovations stemming from exciting theoretical developments [6], [7] but which advocate psychological approaches which promote the suppression of memory for traumatic experiences as way of dealing with negative sequelae, since suppression is clinically contra-indicated [8], [9].

The psychological intervention with the strongest evidence-base for full-blown PTSD is trauma-focussed Cognitive Behaviour Therapy–a treatment which is only indicated when delivered weeks or months after the trauma [10]. However, what can be given to trauma victims suffering flashbacks in the first few weeks? Unfortunately, using talking therapy as a crisis intervention in the immediate aftermath of trauma has caused international clinical concern [11]: interventions such as critical incidence stress debriefing can worsen rather than ameliorate later trauma symptoms [12]. Given the scale of traumatic events globally–war, terrorism, natural disasters, interpersonal violence–there is a huge unmet need for widely-available and easily accessible interventions. We need to develop fresh theory-driven interventions to reduce the build up of flashbacks in the early post-trauma period.

Can cognitive science suggest a way to reduce the build-up of flashbacks to trauma? We suggest that certain cognitive tasks, informed by the neuropsychological domain of working memory, may indeed be used tap into processes underlying flashback memory consolidation. This is based on two key findings: first, cognitive science has shown the brain has selective resources with limited capacity [13]; second, the neurobiology of memory suggests there is a 6-hr window to disrupt memory consolidation [14], [15]. In particular, we suggest that visuospatial tasks will be useful in this regard according to the following rationale: (1) trauma flashbacks are sensory-perceptual images with visuospatial components [2], [16], [17]. (2) visuospatial cognitive tasks compete for resources with visuospatial images [18]–[21]. (3) the neurobiology of memory consolidation suggests a 6-hr time frame post-event within which memories are malleable [14], [15]. Thus (4) we predict visuospatial cognitive tasks given within 6-hr post-trauma will reduce flashbacks.

Following a long tradition in experimental psychopathology, we use the trauma film paradigm [22] in our laboratory as an experimental analog of viewing real trauma and the subsequent flashbacks [23], [24] suffered in PTSD. We have previously demonstrated that the frequency of flashbacks for analog traumatic content can be manipulated experimentally by completing standardized cognitive tasks during trauma film viewing (i.e. peri-traumatically): In healthy participants, completing a complex visuospatial working memory task [19], such as finger pattern tapping, during exposure to a trauma film subsequently reduced flashbacks of that film over 1-week compared to no task, or non-visuospatial tapping [24], [25]. Interestingly, and in line with cognitive theories of PTSD [2], verbal cognitive tasks during the film, such as counting backwards in threes, actually increased the number of flashbacks, confirming that the effects were not simply due to distraction for general working memory resources but to the specific nature of the task [24].

In the current study it was critical to identify a visuospatial working memory task that would be widely-available in the real-world. Clearly a standard neuropsychological test battery visuospatial task such as the WAIS block design [26] would be impractical to deliver en mass. The popular computer game "Tetris" has been demonstrated to be a visuospatial task [27]–[30], drawing on mental rotation and the type of processing we recruit when forming mental images. Playing "Tetris" can even result in participants experiencing subsequent visual images of the game itself at a later time [30], implicating its involvement with intrusive, image-based memory. The capacity of visual memory is both limited and vulnerable to proactive interference, i.e. interruption of memory for presented stimuli by the presentation of similar, but different stimuli after a time delay [31]. Thus "Tetris" provides a promising candidate. Interestingly, watching violent films and playing prolonged computer games are typically associated with negative impact on psychological well being and behavior in both children and adults [32]. This has led to public concern over their ready availability. However, clearly not all computer games are bad for you [27].

We now report the impact of a cognitive visuospatial task, a computerized mental rotation game, on the modulation of analog flashbacks to trauma. Given our earlier experimental demonstration that engaging in visuospatial tasks peri-traumatically (i.e. during trauma film viewing) reduces subsequent intrusive imagery (analog flashbacks), we tested here whether the window for intervention could be extended into the post-trauma period (i.e. after trauma film viewing). This is a critical question since in real-world trauma successful manipulation of flashbacks would need to be conducted post-event rather than peri-traumatically. Further, investigating a time-frame for intervention in the near aftermath of trauma, yet within a window of memory malleability has clear clinical implications. Recent statistics indicate that the average waiting time in an emergency department in the United States is 30-mins [33]. To test the real-world application of our paradigm, we explored the manipulation of flashbacks in the laboratory 30-min after watching a trauma film.

Our suggestion to manipulate flashbacks in the aftermath of a traumatic experience is supported by various experimental demonstrations that newly formed memories are initially labile for a short time and thus subject to interference. In rats, the consolidation of fear memories can be inhibited in a dose dependent fashion with administration of either anisomycin or Rp-cAMPS in the immediate post-training phase [34]. Furthermore, even established memories may be subject to manipulation upon reactivation–the triggering of previously consolidated memories may return them to a labile state and in need of reconsolidation if they are to persist [14], [35], [36]. In humans, the consolidation or stabilization of motor memory (in this case, a finger pattern tapping task) has been demonstrated to occur in the first 6-hours following initial learning: learning a new variation on the motor task after reactivation of the previously learned motor task can block the reconsolidation of memory for the original task [15]. Although there are doubts as to whether such reconsolidation may occur for all types of memory [37], of particular interest to PTSD research is that flashbacks for trauma may be pharmacologically modulated [4], [38], though this has not been without controversy [5]. Given the potential advantages of a non-invasive procedure, we were interested in testing whether a cognitive intervention-a visuospatial task-could modulate later flashbacks to traumatic material.

We predicted that playing a visuospatial computer game requiring mental rotation of shapes ("Tetris" [39]), 30-min after viewing graphic and traumatic film footage, would help reduce later involuntary flashbacks of the traumatic material, but leave voluntary memory retrieval intact. Similarly we predicted that playing "Tetris" would be associated with reduced clinical symptomatology at one week.

Results

Forty participants watched a 12-min film of traumatic scenes of injury and death (n = 20 per group). Film viewing was followed by a 30-min interval before simple random assignment to one of two experimental conditions (Figure 1). There were no baseline differences between the two groups in terms of age, depressive symptoms or trait anxiety (Table 1) or gender. Mood was equivalent between the groups prior to watching the film, and as predicted, both groups experienced comparable mood deterioration following the film (Table 2).

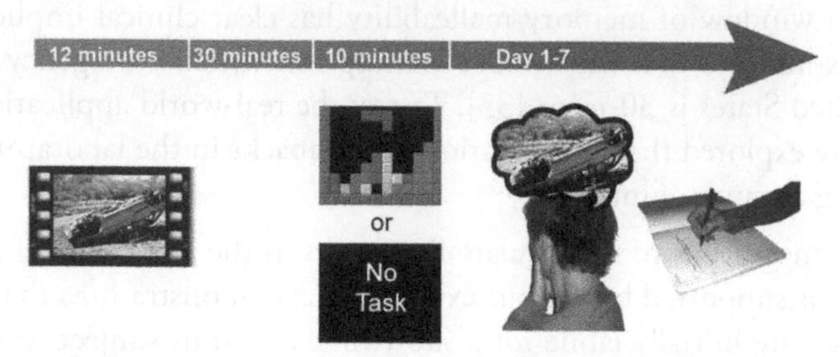

Figure 1. Study design overview. Participants completed a well-established experimental analog for PTSD, the trauma film paradigm. All participants viewed a traumatic film consisting of scenes of real injury and death and had a 30-min structured break. Participants were randomly allocated to either a no-task or visuospatial condition which they undertook for 10-min. Flashbacks (involuntary memories) were monitored for 1-week using a daily diary. Then participants returned to the laboratory for an assessment of clinical symptomatology relating to the flashbacks as well as a test of voluntary memory (recognition memory).

Table 1. Means and statistics for age and baseline assessments indicating experimental groups were equivalent at baseline.

Measure	Visuospatial (n = 20)		No-task (n = 20)		t-test
	mean	sem	mean	sem	
Age	22.4	1.4	24.4	1.1	$t_{(38)} = 1.15$ (NS)
Beck Depression Inventory	6.2	1.2	5.1	1.6	$t_{(38)} = 0.56$ (NS)
Trait Anxiety (STAI-T)	39.3	2.2	36.5	1.7	$t_{(38)} = 1.00$ (NS)

Table 2. Means and statistics for negative mood assessment (pre vs post-trauma film) indicating equivalent deterioration in mood across conditions prior to the administration of the experimental task.

	Visuospatial (n = 20)		No-task (n = 20)		ANOVA		
	mean	sem	mean	sem	Time	Group	Group*Time
Pre-Trauma Film Mood	5.1	1.0	4.3	0.8	$F_{(1, 38)} = 26.81$ [$]	$F_{(1, 38)} = 0.06$ (NS)	$F_{(1, 38)} = 1.76$ (NS)
Post-Trauma Film Mood	8.4	1.2	9.9	1.3			

[$] $p < 0.01$.

Following the 30-min interval period in which participants completed standardized filler tasks, a brief reminder task for the trauma film was administered to both groups. Participants then either completed the visuospatial condition or sat quietly (no-task control condition) for 10-min. During these 10-min, all participants recorded the frequency of trauma film flashbacks they experienced. Significantly fewer flashbacks were experienced in this initial period (i.e. while playing the game or not) during the visuospatial condition than the no-task condition (Table 3).

Table 3. Frequency of initial flashbacks during the 10-min experimental task indicating that there were significantly fewer involuntary flashback memories during the visuospatial compared to the no-task control condition.

Group	mean	sem	t-test
Visuospatial (n = 20)	4.6	1.1	$t_{(38)} = 2.50^{\dagger}$
No-task (n = 20)	12.8	3.1	

† = p<0.05.

After leaving the laboratory, participants then kept a daily diary in which they recorded their flashbacks to the trauma film over a period of 1-week. Crucially, we found that participants in the visuospatial condition experienced significantly fewer flashbacks over the week (Figure 2) than those in the control condition. Furthermore, at 1-week, participants returned to the laboratory–participants in the game condition had significantly lower scores on the measure of clinical symptomatology of trauma–the Impact of Events Scale [40] (Figure 3).

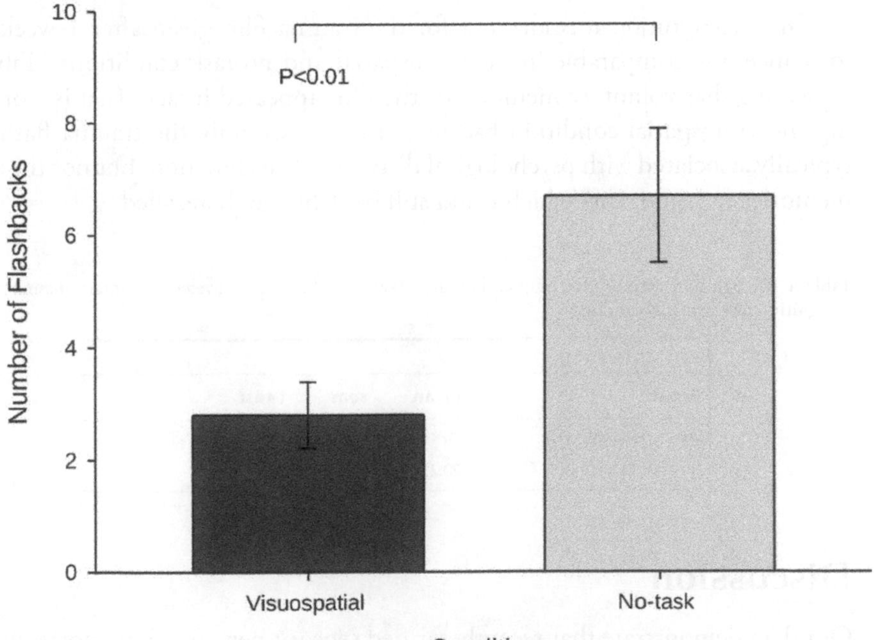

Figure 2. Flashback frequency over 1- week. As predicted, there was a significant reduction in the number of flashbacks over 1-week in the visuospatial condition compared to no-task condition, t(38) = 2.87 (mean+/–sem).

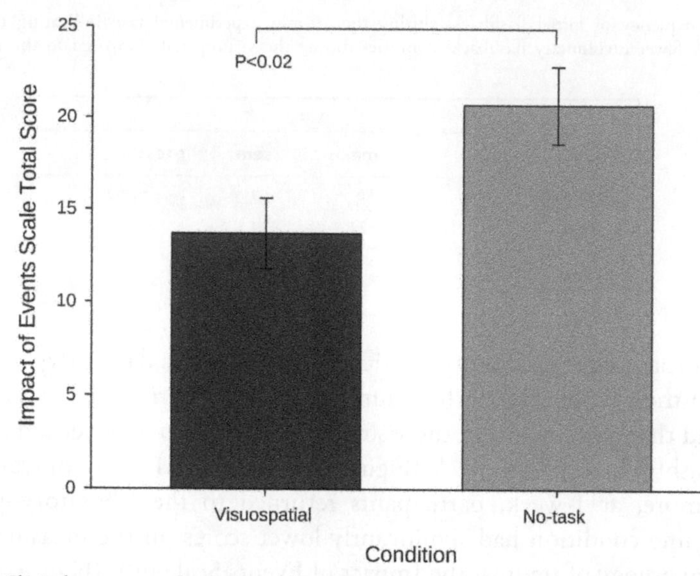

Figure 3. Clinical symptomatology at 1-week. Impact of Event Scale scores at 1-week indicated significantly lower impairment in the visuospatial condition compared to no-task condition, t(38) = 2.47 (mean+/−sem).

On a recognition memory test for the trauma film given after 1-week, performance was comparable in the visuospatial and no-task conditions (Table 4), indicating that voluntary memory for the film appeared intact. That is, completing the visuospatial condition had served to reduce only the trauma flashbacks typically associated with psychological distress and dysfunction, but not the actual memories for the events which could still be deliberately recalled.

Table 4. Recognition memory for trauma film after 1-week indicating equivalent voluntary memory on the recognition task across conditions.

Group	mean	sem	t-test
Visuospatial (n = 20)	20.1	0.8	$t_{(38)} = 0.10$, (NS)
No-task (n = 20)	20.2	0.7	

Discussion

Our data demonstrate that recently formed sensory-perceptual memories are vulnerable to manipulation 30-min following watching a traumatic film. Why more precisely might visuospatial computer games be effective in reducing at least analog flashbacks following trauma? In accordance with the longstanding psychological model of human memory–the working memory model [13], [18]–[20], [41], [42], we propose that strategic, selective interference with the consolidation of recently triggered visual memories occurs via the demand on the player's limited visuospatial working memory resources. The major clinical theories of PTSD [2], [43], [44], [45] converge to suggest that there are two forms of processing that occur simultaneously for any given traumatic event: (1) the sensory-perceptual processing of the trauma e.g. the sights and sounds experienced during a car crash; (2) verbal or conceptual processing e.g. making sense or a coherent narrative about what is occurring. It is information from sensory-perceptual processing that provides the foundation for flashback images. Clinical models of PTSD propose that the relative balance of sensory-perceptual versus verbal/conceptual processing of a traumatic event determines whether flashbacks are formed, whereby a skewed balance towards sensory-perceptual aspects of the trauma is pathological.

When viewing traumatic film stimuli, the type of sensory-perceptual focus engaged in is predominantly visual (rather than say olfactory or auditory). To interfere with this type of visual processing we need to target what are known cognitively as 'visuospatial' resources [19]. That is, visuospatial tasks that use the same type of processing as do visual flashbacks will interrupt memory consolidation of those flashbacks by competition for the same limited cognitive resources [24]. Thus, by selectively interfering with visual sensory-perceptual processing of the traumatic film via visuospatially demanding cognitive tasks, subsequent analog flashbacks are reduced. Note, this is not the same as simple distraction, since other types of tasks such as verbal tasks during traumatic films are predicted (and have been shown) to lead to increased flashbacks [23], [24]. Our data is the first indication that the manipulation of visuospatial processing in the consolidation phase of recently activated trauma memories can serve to modulate future intrusive, involuntary flashbacks (despite leaving voluntary memory intact). "Tetris" participants experience fewer intrusions even while playing the game, supporting the competition for resources rationale. Significantly, we demonstrate that the visuospatial task conducted 30-min post-exposure to traumatic stimuli is effective in reducing flashbacks of that trauma as well as associated clinical symptomatology over 1-week.

Interestingly, the clinical literature offers potentially converging support for these findings. Eye Movement Densensitization and Reprocessing (EMDR) is an empirically-supported treatment for established PTSD [10]. During this therapy, the patient undergoes a series of eye movements whilst holding an intrusive traumatic memory in mind, leading to a reduction in the emotionality and vividness of the unpleasant mental image. One of several possible accounts of how EMDR might work is that the eye movements draw on visuospatial processing [18], [46], [47] and thus provide a dual task competing specifically for resources with the trauma imagery, reducing its impact. Our proposal similarly draws on a working memory rationale for how "Tetris" may affect flashback formation in a modality specific manner [48], though emphasizes all the senses involved in imagery (not just the visual modality). However, a critical difference between the current experiment and EMDR is that EMDR is used for treating existing flashbacks in PTSD (at least one month post-trauma), but is not intended to be used during the memory consolidation phase targeted in the current study. Our interest in the immediate aftermath of trauma is to understand preventative (rather than just curative) measures to the development of PTSD flashbacks.

Speculatively, the effects of "Tetris" may not be limited to the immediate post-trauma period during which it is played but may even continue to compete for visuospatial resources later. For example, it has already been demonstrated [30] that images of "Tetris" can intrude during sleep-a period during which memory consolidation is known to occur. Future research is required to examine the precise mechanisms of action by which "Tetris" reduces flashbacks to trauma. We predict that a verbal task would not have comparable effects and may even worsen flashbacks. Thus future studies should compare both a visuospatial task (e.g. "Tetris") with a verbal task (e.g. a verbal computer game) against a no-task control group.

Our alternative and novel approach of using cognitive visuospatial tasks, rather than pharmacological means to reduce flashbacks following trauma aims to deal directly with the consolidation and potentially, reconsolidation, of such intrusive imagery in an ethical, safe and economical way. We suggest this approach could be harnessed as a 'cognitive vaccine' to inoculate against escalation of flashbacks contributing to full blown PTSD. Further research is required but potential clinical applications of our paradigm include use by emergency services in the early post-trauma period, e.g. to victims of rape or delivering such tasks to populations subject to regular trauma exposure e.g. firefighters or those involved in armed combat. To better map the horizons of human memory, we should further delineate the clinical possibilities offered by cognitive theory to reduce pathological aspects of memory, such as flashbacks.

Materials and Methods

Forty participants (aged 18–47 years; mean age = 23 years; 22 males) completed baseline assessments of mood, trait anxiety and depression and then viewed traumatic film footage (Figure 1). The 12-min film contained 11 clips of traumatic content including graphic real scenes of human surgery, fatal road traffic accidents and drowning. Following the film, mood assessments were repeated and standardized filler tasks completed for 30-min. A brief film reminder task was then administered in which one neutral but recognizable image from each film clip was presented to all participants.

According to randomization to condition (simple, computer-generated random allocation), participants either completed the visuospatial condition or were in a no-task control condition for 10-min. Participants in the visuospatial condition played the game ("Tetris") on a computer and used the cursor keys to move and rotate falling blocks to complete the largest number of complete rows across the screen. In the no-task control condition, participants were asked to sit quietly for 10-min. During both conditions participants recorded initial flashbacks of the trauma over the 10-min.

Participants then kept a daily diary for 1-week, in which they recorded and described (for verification) each of their flashbacks, i.e., spontaneously occurring image-based intrusions of the trauma film (based on [24]). On return to the laboratory 1-week later, participants completed a recognition memory task to index voluntary memory retrieval. The task comprised a series of 32 written statements regarding the film, presented individually (e.g. 'Three cars were involved in the crash'). Participants responded to each statement with either "true" or "false" and scored one point for each correct response. Finally, participants completed the Impact of Events Scale [40]–a clinical measure of PTSD symptomatic response adapted to the trauma film over the past 1-week.

Acknowledgements

The authors would like to thank Mike Berger and Corin Bourne for thoughtful discussion.

Authors' Contributions

Conceived and designed the experiments: EAH. Performed the experiments: EAH ELJ TCB. Analyzed the data: EAH ELJ TCB CD. Contributed reagents/materials/analysis tools: EAH. Wrote the paper: EAH ELJ TCB CD.

References

1. American Psychiatric Association (2000) Diagnostic and Statistical Manual of Mental Disorders. Washington D.C.: American Psychiatric Association.
2. Brewin CR, Holmes EA (2003) Psychological theories of posttraumatic stress disorder. Clinical Psychology Review 23: 339–376.

3. Bryant RA, Harvey AG (2000) Avoidant coping style and post-traumatic stress following motor vehicle accidents. Behaviour Research and Therapy 33: 631–635.
4. Pitman RK, Sanders KM, Zusman RM, Healy AR, Cheema F, et al. (2002) Pilot Study of Secondary Prevention of Posttraumatic Stress Disorder with Propranolol. Biological Psychiatry 51: 189–192.
5. Henry R, Fishman JR, Youngner SJ (2007) Propranolol and the prevention of post-traumatic stress disorder: Is it wrong to erase the "sting" of bad memories? The American Journal of Bioethics 7: 12–20.
6. Anderson MC, Green C (2001) Suppressing unwanted memories by executive control. Nature 410: 366–369.
7. Hertel PT, Calcaterra G (2005) Intentional forgetting benefits from thought substitution. Psychonomic Bulletin & Review 12: 484–489.
8. Depue BE, Curran T, Banich MT (2007) Prefrontal regions orchestrate suppression of emotional memories via a two-phase process. Science 317: 215–219.
9. Holmes EA, Moulds ML, Kavanagh D (2007) Memory suppression in PTSD treatment [Letter to the Editor]. Science 722.
10. National Institute for Health and Clinical Excellence (2005) Post-traumatic stress disorder (PTSD): the management of PTSD in adults and children in primary and secondary care. London National Institute for Health and Clinical Excellence. CG026.
11. McNally RJ, Bryant RA, Ehlers A (2003) Does early psychological intervention promote recovery from posttraumatic stress? Psychological Science in the Public Interest 4: 45–79.
12. Mayou RA, Ehlers A, Hobbs M (2000) Psychological debriefing for road traffic accident victims: Three-year follow-up of a randomized controlled trial. British Journal of Psychiatry 176: 589–593.
13. Baddeley AD (2003) Working memory: looking back and looking forward. Nature Reviews Neuroscience 4: 829–839.
14. Nader K (2003) Memory traces unbound. Trends in Neurosciences 26: 65–72.
15. Walker MP, Brakefield T, Hobson JA, Stickgold R (2003) Dissociable stages of human memory consolidation and reconsolidation. Nature 425: 616–620.
16. Ehlers A, Hackmann A, Michael T (2004) Intrusive re-experiencing in post-traumatic stress disorder: Phenomenology, theory, and therapy. Memory 12: 403–415.

17. Holmes EA, Grey N, Young KAD (2005) Intrusive images and "hotspots" of trauma memories in posttraumatic stress disorder: An exploratory investigation of emotions and cognitive themes. Journal of Behavior Therapy and Experimental Psychiatry 36: 3–17.
18. Andrade J, Kavanagh DJ, Baddeley A (1997) Eye-movements and visual imagery: A working memory approach to the treatment of post-traumatic stress disorder. British Journal of Clinical Psychology 36: 209–223.
19. Baddeley AD, Andrade J (2000) Working memory and the vividness of imagery. Journal of Experimental Psychology-General 129: 126–145.
20. Kavanagh DJ, Freese S, Andrade J, May J (2001) Effects of visuospatial tasks on desensitization to emotive memories. British Journal of Clinical Psychology 40: 267–280.
21. van den Hout M, Muris P, Salemink E, Kindt M (2001) Autobiographical memories become less vivid and emotional after eye movements. British Journal of Clinical Psychology 36: 209–223.
22. Horowitz MJ (1969) Psychic trauma. Return of images after a stressful film. Archives of General Psychiatry 20: 552–559.
23. Holmes EA, Bourne C (2008) Inducing and modulating intrusive emotional memories: A review of the trauma film paradigm. Acta Psychologica 127: 553–566.
24. Holmes EA, Brewin CR, Hennessy RG (2004) Trauma films, information processing, and intrusive memory development. Journal of Experimental Psychology: General 133: 3–22.
25. Stuart ADP, Holmes EA, Brewin CR (2006) The influence of a visuospatial grounding task on intrusive images of a traumatic film. Behaviour Research and Therapy 44: 611–619.
26. Weschsler D (1981) WAIS-R manual. New York: The Psychological Corporation.
27. Green SC, Bavelier D (2003) Action video game modifies visual selective attention [letter to the editor]. Nature 423: 534–537.
28. Haier RJ, Siegel BV, MacLachlan A, Soderling E, Lottenberg S, et al. (1992) Regional glucose metabolic changes after learning a complex visuospatial/motor task: a positron emission tomographic study. Brain Research 570: 134–143.
29. Sims VK, Mayer RE (2002) Domain specificity of spatial expertise: The case of players. Applied Cognitive Psychology 16: 97–115.

30. Stickgold R, Malia A, Maguire D, Roddenbury D, O'Connor M (2000) Replaying the game: Hypnagogic images in normals and amnesics. Science 290: 350–353.

31. Hartshorne JK (2008) Visual working memory capacity and proactive interference. PLoS ONE 3: e2716.

32. Anderson CA, Gentile DA, Buckley KE (2007) Violent Video Game Effects on Children and Adolescents: Theory, Research and Public Policy. Oxford: Oxford University Press.

33. Wilper AP, Woolhandler S, Lasser KE, McCormick D, Cutrona SL, et al. (2008) Waits to see an emergency department physician: U.S. trends and predictors, 1997–2004. Health Affairs 27: w84–w95.

34. Schafe GE, LeDoux JE (2000) Memory consolidation of auditory pavlovian fear conditioning requires protein synthesis and protein kinase A in the amygdala. The Journal of Neuroscience 20: RC96.

35. Rudy JW (2008) Destroying memories to strengthen them. Nature Neuroscience 11: 1241–1242.

36. Lee JLC (2008) Memory reconsolidation mediates the strengthening of memories by additional learning. Nature Neuroscience 11: 1264–1266.

37. Dudai Y (2006) Reconsolidation: the advantage of being refocused. Current Opinion in Neurobiology 16: 174–178.

38. Brunet A, Orr SP, Tremblay J, Robertson K, Nader K, et al. (2008) Effect of post-retrieval propranolol on psychophysiologic responding during subsequent script-driven traumatic imagery in post-traumatic stress disorder. Journal of Psychiatric Research 42: 503–506.

39. The PC version of Tetris (copyright 1985–2008, Tetris Holdings LLC.) was obtained from Blue Planet Software (Honolulu, HI).

40. Horowitz M, Wilner N, Alvarez W (1979) Impact of event scale: A measure of subjective stress. Psychosomatic Medicine 41: 209–218.

41. Kavanagh DJ, Andrade J, May J (2005) Imaginary relish and exquisite torture: The elaborated intrusion theory of desire. Psychological Review 112: 446–467.

42. Kemps E, Tiggerman M, Woods D, Soekov B (2004) Reduction of food cravings through concurrent visuospatial processing. International Journal of Eating Disorders 36: 31–40.

43. Conway MA, Pleydell-Pearce CW (2000) The construction of autobiographical memories in the self- memory system. Psychological Review 107: 261–288.

44. Dalgleish T (2004) Cognitive approaches to posttraumatic stress disorder: the evolution of multirepresentational theorizing. Psychological Bulletin 130: 228–260.

45. Ehlers A, Clark DM (2000) A cognitive model of posttraumatic stress disorder. Behaviour Research and Therapy 38: 319–345.

46. Gunter RW, Bodner GE (2008) How eye movements affect unpleasant memories: Support for a working memory account. Behaviour Research and Therapy 46: 913–931.

47. Lilley SA, Andrade J, Turpin G, Sabin-Farrell R, Holmes EA (in press) Visuospatial working memory interferes with recollections of trauma. British Journal of Clinical Psychology.

48. Kemps E, Tiggemann M (2007) Modality-specific imagery reduces cravings for food: an application of the elaborated intrusion theory of desire to food craving. Journal of Experimental Psychology: Applied 13: 95–104.

Unofficial Policy: Access to Housing, Housing Information and Social Services Among Homeless Drug Users in Hartford, Connecticut

Julia Dickson-Gomez, Mark Convey, Helena Hilario, A. Michelle Corbett and Margaret Weeks

ABSTRACT

Background

Much research has shown that the homeless have higher rates of substance abuse problems than housed populations and that substance abuse increases individuals' vulnerability to homelessness. However, the effects of housing policies on drug users' access to housing have been understudied to date. This paper will look at the "unofficial" housing policies that affect drug users' access to housing.

Methods

Qualitative interviews were conducted with 65 active users of heroin and cocaine at baseline, 3 and 6 months. Participants were purposively sampled to reflect a variety of housing statuses including homeless on the streets, in shelters, "doubled-up" with family or friends, or permanently housed in subsidized, unsubsidized or supportive housing. Key informant interviews and two focus group interviews were conducted with 15 housing caseworkers. Data were analyzed to explore the processes by which drug users receive information about different housing subsidies and welfare benefits, and their experiences in applying for these.

Results

A number of unofficial policy mechanisms limit drug users' access to housing, information and services, including limited outreach to non-shelter using homeless regarding housing programs, service provider priorities, and service provider discretion in processing applications and providing services.

Conclusion

Unofficial policy, i.e. the mechanisms used by caseworkers to ration scarce housing resources, is as important as official housing policies in limiting drug users' access to housing. Drug users' descriptions of their experiences working with caseworkers to obtain permanent, affordable housing, provide insights as to how access to supportive and subsidized housing can be improved for this population.

Background

Researchers studying the causes of homelessness have frequently engaged in a polarized debate. Many have looked to the personal factors of homeless individuals as causes of homelessness [1,2]. One personal factor that has been hypothesized as a cause of homelessness is drug abuse. Research has shown that substance use problems afflict anywhere from 28 to 67% of homeless individuals [3-7] and that substance abuse increases individuals' vulnerability to homelessness [8-10]. Others have argued that structural changes, for example the loss of manufacturing jobs and affordable housing stock in inner-city neighborhoods, are the causes of the increase in homelessness over the past two decades [11,12]. More recently researchers have argued that both are important considerations. While personal characteristics, such as drug use, may not in themselves cause homelessness, they make certain individuals more vulnerable to homelessness given an increasingly competitive housing market [13-18]. Structural factors determine why pervasive homelessness exists in this historical time, while individual factors explain who is least able to compete for scarce affordable housing.

Structural factors that may contribute to drug users' greater vulnerability to homelessness include official and unofficial housing policies that determine eligibility for and access to various housing and welfare subsidies. The effects of housing policies on drug users' access to housing have been understudied to date. Official policies include the federal "One Strike and You're Out" law (P.L. 104–120, Sec.9) passed in 1996 that allows federal housing authorities to consider drug and alcohol abuse and convictions of people and their family members when making decisions to evict them from or deny access to federally subsidized housing. Many states, including Connecticut, have opted out of this law. Flat line funding of federally subsidized housing programs, such as the Housing Choice voucher program (formerly known as Section 8), and Shelter Plus Care, have limited the number of subsidies available. Both programs allow recipients to choose their own apartments on the competitive market and pay a proportion of the rent depending on recipients' income. While Connecticut does not consider drug convictions in decisions to deny applications for housing vouchers, criminalization of drug use affects drug users' access to housing in other ways, as criminal background checks are routine in many apartment rental applications. Other policies which have impacted drug users' access to housing include the Personal Responsibility and Work Opportunity Act of 1996, popularly known as Welfare Reform, in particular the elimination of the SSI Addiction Disability and a ban on receiving welfare benefits for convicted drug offenders [19-21].

Less understood are the effects of "unofficial" policy on drug users' access to housing. In this paper, unofficial policy is defined as the way in which official policy is implemented or enforced, or not, and the operating policies of organizations or individuals. This definition of unofficial policy borrows from Lipsky's [22] idea of "street level bureaucrats." For Lipsky, low-level employees who directly interact with the public, for example social workers, police officers, or unemployment counselors, ought to be viewed as policy makers rather than implementers of policy. As Lipsky puts it, the "decisions of street level bureaucrats, the routines they establish, and the devices they invent to cope with pressure, effectively become the public policy they carry out (xxii)." The pressures of work faced by street level bureaucrats include an almost infinite demand for services by the public along with inadequate resources available to workers to meet these demands. Street level bureaucrats use a number of strategies to ration services, including limiting access to information about services, creating categories of clients, exercising discretion in distributing benefits and sanctions, and increasing the costs of applying for services. Lipsky does not fully consider, however, the ways official policy may shape unofficial policy. For example, an official policy that cuts federally subsidized housing may create periods of relative scarcity, which may have direct effects on the pressures and coping mechanisms street level bureaucrats use.

Unofficial policy may help explain research that has shown that substance users are significantly less likely to exit homelessness [23] or access social services [24,25] than non-substance abusing homeless. For example, Zlotnick [23] and colleagues found that exit from homelessness was associated with greater social support and greater contact with service providers for homeless without a current substance abuse disorder, but not for homeless with current substance abuse. They suggest that this may be because substance using homeless persons may be more focused on obtaining and using drugs than gaining access to services, or that they may be unable to mobilize their social support networks. An alternative explanation, consistent with Lipsky's view of "street level bureaucrats" is that service providers may choose to devote more of their limited resources to homeless individuals without substance abuse problems whom they may see as more "deserving" or as having a greater chance at success in maintaining their housing. Supporting this second explanation is a study by Dohan and colleagues [26] that found that welfare workers generally applauded welfare reform's renewed attention to deservingness, including program emphases on client self-sufficiency and personal accountability.

Some drug users face multiple barriers to accessing and maintaining stable housing, including long-term substance abuse, mental health issues, and histories of arrest. Such individuals have been identified by researchers and advocates as "chronically homeless" [27-30]. As a result, alternatives to emergency shelters to house this population have begun to be proposed, including the Housing First Model, and supportive housing programs [30]. The Housing First model advocates for the provision of housing to drug addicted or mentally ill homeless that is not contingent on their "readiness," i.e., completing residential treatment programs or maintaining sobriety for a period of time. Rather, they advocate for housing with supportive services attached, including mental health services, addiction services, and assistance in budgeting, obtaining employment or maintaining an apartment. This is in contrast to the traditional Continuum of Care model that consists of several components including outreach, treatment and transitional housing, then supportive housing. Continuum of Care seeks to enhance clients' "housing readiness" by requiring sobriety and compliance with psychiatric treatment before placement to more permanent housing [30]. Connecticut has funded several supportive housing projects that provide affordable, service enriched rental housing for homeless and at-risk populations, many of whom are coping with mental illness, histories of substance addiction, or HIV/AIDS [31]. Some of these supportive housing programs follow the Housing First model and allow residents to choose which, if any, supportive services they wish to utilize. Other programs require residents to fulfill program requirements, such as active involvement in job training or substance abuse treatment. The effects these differing philosophies have on the ways in which service providers implement programs, i.e. the unofficial policy of these programs, has not been studied.

This paper will look at the "unofficial" policies that affect drug users' access to housing. Using longitudinal, in-depth interviews with both housed and homeless drug users and key informant interviews with housing caseworkers in Hartford, Connecticut, we will look at the process by which drug users receive information about different housing subsidies and welfare benefits, and their experiences in applying for these.

Methods

Design

We conducted longitudinal in-depth interviews with active drug users to explore their housing status and stability over time, and barriers and facilitators drug users face in accessing housing. Eligibility criteria included being over 18 years old and having used cocaine, crack or heroin within the last 30 days at the first interview. We sought to recruit active users of heroin and cocaine because previous research conducted by our research team indicated that these were the illicit drugs most frequently abused in Hartford, and that users of these substances had steadily increasing rates of homelessness over the past thirteen years [32-34]. Purposive sampling was used to identify and recruit drug users in various housing situations, including: 1) supportive housing, 2) subsidized housing, 3) non-subsidized housing, 4)"doubling up" with family or friends, 5) homeless in shelters, and 6) homeless on the street. We defined "doubling up" as the practice of temporarily moving in with family or friends.

In addition, we conducted key informant interviews and focus group interviews with service providers including shelter, supportive housing, and substance abuse treatment staff, and housing advocates in order to obtain service provider perspectives on the barriers and facilitators drug users face in accessing information, housing and services.

Participants

Sixty-five drug users were interviewed at baseline. Forty-six percent of the sample was African American, 46% Puerto Rican, 8% non-Hispanic white, and 46%

women. Participants were ethnically similar to other research projects conducted with active drug users in Hartford, although women were oversampled [33,35]. Fifty were located for follow-up interviews at three months. Of those who were not located, four were confirmed to be in jail, and one was confirmed to have moved out of state. Excluding those individuals who were in jail or had moved results in an overall retention rate of 83%. Forty-one were located for interviews at 6 months. Of those who were not located at 6 months, two were deceased, two were confirmed to have moved out of state, and five were in jail. Excluding those who had died, gone to jail, or moved out of state resulted in an overall retention rate of 73.2%. The refusal rate was less than 5%.

We conducted key informant interviews with six service providers including staff at three area shelters, leaders of groups advocating for low-income housing or to end homelessness, and staff at a substance abuse treatment organization. Two focus group interviews with three and four participants each were conducted with staff from an additional shelter and staff at an organization administering several supportive housing programs. These were originally designed and intended to be key informant interviews. However, staff at each organization expressed interest in being interviewed together so that they could share and compare their perspectives and experiences. Key informants and focus group participants included staff in different positions within their organizations, including the executive director of one organization, supervisors and caseworkers with more direct, daily interactions with clients. Participants were 60% female, 60% white, 30% African American, and 10% Latino. The refusal rate among service providers was approximately 50%. Most refusals were due to time constraints or scheduling problems.

Procedure

Participant recruitment for the drug using sample was achieved through a combination of direct street recruitment and referral from other projects. For participants who were directly recruited, we targeted recruitment in locations where populations of drug users with differing housing characteristics could be found. Drug users who were homeless were recruited from each of Hartford's seven shelters or soup kitchens. Outreach staff approached potential participants in these settings, distributed HIV prevention materials such as bleach kits and condoms to initiate a general discussion about risk behaviors and assess their general eligibility for the study. Those participants who appeared interested and eligible were given an appointment card for full screening. Drug users who were doubled up with family or friends or housed in subsidized, non-subsidized or supportive housing were similarly recruited through street outreach, or from prior knowledge of their situation from ethnographic research in other research projects working with active drug users. We attempted to recruit equal numbers of drug users (approximately 10 or 11) from each of the housing statuses. In practice it was much easier to recruit participants in some housing categories than others (e.g. homeless in shelter and participants doubled up with family or friends were easier to identify and recruit than homeless on the street or drug users in supportive housing). Therefore, throughout the course of recruitment, when participants in any particular housing category became overrepresented, recruitment for that housing category was stopped and outreach and recruitment efforts focused on finding drug users in under-represented housing situations. Table 1 shows the housing status of participants at baseline, 3 and 6 months. Participants received a $25 incentive for completing each interview and a $15 bonus for completing all three interviews. Interviews were approximately 1 1/2 hours in duration. Written informed consent was obtained from all participants, both drug users and service providers, and the research protocol was approved by the Institutional Review Board at the Institute for Community Research.

Table 1. Frequency of Housing Statuses among Active Drug Users

	Baseline (N = 65)	3-month (N = 50)	6-month (N = 40)
Shelter	17%	28%	18%
On the street	14%	10%	5%
Double up	31%	22%	31%
Supportive	6%	6%	10%
Subsidized	15%	16%	26%
Non-subsidized	17%	18%	10%

All in-depth interview guides were project developed. Baseline interviews with drug users explored participants' housing histories over the previous two years, focusing on: reasons for moves, evictions or housing changes; types of public assistance, social services and housing subsidies applied for and accessed; the amount of time elapsed between application for housing and other social services and receipt or denial of housing or other services; and reasons given for denial of housing programs or apartment applications. To help participants construct their housing histories, we asked them to describe their current living situations and then moved back in time.

Three month follow-up interviews explored changes in housing status and access to housing programs. Baseline interviews were reviewed prior to follow-up interviews so that interview questions could be focused on participants' specific situations. If housing status changed since baseline, interviewers explored reasons for moves, eviction or housing changes, and any new applications to public assistance, social services or housing subsidies. The status or outcome of applications made or planned at baseline were explored. Six month follow-up interviews used the same interview guide as three-month interviews. Again, three-month interviews were reviewed so that questions followed up on any housing changes planned or made. Three-month interviews also included a brief quantitative survey to collect basic demographic information including age, income, length of time living in Hartford, educational level, and quantity and frequency of use of a variety of different drugs. This brief survey was added after it was determined that it was difficult to quantify such information from qualitative interviews. Six-month interviews also included brief demographic surveys that collected information on income, quantity and frequency of drug use in the last 30 days.

Service providers for key informant and focus group interviews were selected to represent a variety of organizations that may be directly or indirectly involved in assisting drug users to obtain housing. A list that included local homeless shelters, soup kitchens, drug treatment centers, mental health organizations, housing and homeless advocacy groups, and supportive housing programs, was compiled from staff knowledge, internet searches and networking with housing advocates and service providers. Potential staff members to target for interviews were also identified in an attempt to represent the ethnic and professional diversity within organizations. Project ethnographers then directly contacted staff at the organizations, explained the purpose of the study and invited staff to participate in a key informant interview, or contacted supervisors within the organization to explain the purpose of the study and ask permission to contact other staff members to participate in a key informant interview. Written informed consent was obtained from service providers. The length of interviews was 1/2 hour to 45 minutes.

Interviews with service providers focused on the facilitators and barriers that drug users face in accessing independent housing and in maintaining stable housing. We asked service providers about the characteristics of their clientele, including how clients are referred to their organization, to determine initial barriers or facilitators to accessing social or housing services for active drug users. We then asked them to describe the types of housing programs available, other services provided by the organization, and the eligibility requirements for housing and other services for their clients. We also asked them to describe the process through which they try to obtain housing or other services for their clients, the clients whom they have the most difficulty assisting in accessing housing, the strategies, if any, they use to overcome barriers in accessing housing for these difficult clients, and the clients who are the easiest to assist in accessing housing. Finally, we asked providers to describe reasons drug using clients have difficulty maintaining housing and the kinds of support services they feel are necessary to keep drug users in stable housing.

Analysis

All interviews were tape recorded and transcribed verbatim. All text data were coded and analyzed for key themes and patterns of response using Atlas.ti software [36]. Interviews were coded for type of interview (key informant, drug user baseline, three or six month). Interviews with drug users were further coded for demographics and housing status at the time of interview. Data were then coded a first time for content. The coding tree was developed in an iterative process by the research team and applied to in-depth interviews with drug users and key-informant interviews. This first level of analysis coded for broad categories, e.g. social service application process, caseworkers, housing subsidies, shelter, or eviction. After this first level of coding, interviews were coded a second time to further refine categories and emerging themes. For example "creaming", "silting" "costs of applying for services," and "service provider discretion/priorities" were themes that emerged during this second level of coding. Excerpts presented in this paper were chosen to reflect these themes. All names of persons or organizations used in the paper are pseudonyms. Finally, in-depth interview with drug users were analyzed to capture changes over time. After all interviews that a participant had completed were coded, summaries were written for each participant that described his or her housing history, and welfare or other benefits received. Each participant's changes in housing status and the housing subsidies or other benefits applied for or received were then quantified by filling out a Housing Summary Checklist. These data were entered into SPSS and analyzed to show changes in housing status and stability, receipt of welfare or health benefits over time, and associations between housing status and applications to housing programs.

Results

Sample Characteristics

Demographics for participants were collected at three-month interviews. Mean age of participants was 43 years (s.d. 6.8 years). Participants were low income with 63% having earned less than $500 in the last month, and 94% having earned less than $1000. At three months, 54% had smoked crack in the last 30 days, while 30% had injected heroin. Although current drug use was an eligibility criterion at baseline, many participants had entered treatment or stopped drug use by their 3-month interview. Those smoking crack at 3-months smoked a mean of 39 times in the prior 30 days (s.d. 45), while those injecting heroin at 3 months injected an average of 32 times in the prior month (s.d.3.4). Of those who completed all three interviews, 50% had moved at least once during the study period, while 20% had moved 4 or more times indicating a high degree of housing instability.

Shelters as Point of Access to Housing Programs

In comparison with homeless participants who stayed on the street or who doubled up with family members or friends, homeless participants who stayed in shelters reported receiving more information about different housing programs available, particularly supportive housing programs, and were more likely to have applied for these and housing subsidies such as Shelter Plus Care or Section 8. Eight out of the 21 participants who stayed in a shelter at some point during the study period applied for or received supportive housing as compared to 2 out of the 27 participants who were homeless on the street or doubled up with family or friends but had not stayed in a shelter during the study period ($p = .013$). All the area shelters employed caseworkers whose job it is to help shelter residents access more permanent housing and other services, such as mental health or drug treatment. Shelter caseworkers are well informed of new housing programs starting in the city. Some shelters refer clients to programs run by other agencies. Other shelters have started their own supportive housing programs and so can connect their clients directly into their programs as space and funding become available. Shelter residents are often the first to hear of new housing programs, and may be able to apply and receive them before others have even heard about them.

Ethnographer: So what made you eligible to get into the [supportive housing] program at the Horizons Center? Did you have to meet some kind of criteria?

Roger (African American, 53 years old): I was in the shelter first...I had a counselor down there that, you know, we got along real good and she said I'd be a perfect candidate for the Horizons Center. That's when they first was starting the program so they didn't have too many people that knew about it so she sent me over there, you know, and I had to go through some sort of screening.

Supportive housing programs run through the shelter often give preferential treatment to residents of that particular shelter.

Ralph (African American, 36): At the end of the season they [shelter staff] were talking about it [a supportive housing program], telling us about it and nobody really thought it was going to happen. But during the summer...before the [shelter] was going to open up last year, like the outreach workers, some case managers came around to all of the [other] shelters that they were at, that their clients, quote, unquote, their clients would maybe be at or maybe if they just saw us on the street they told us, "On this particular date make sure you come up the back at 9:00 in the morning. You are signing up for housing, first come, first serve..".. And it was a big turnout actually.

The shelter that was offering this particular housing program closed during the summer, so outreach workers visited other area shelters to find residents who had stayed in that shelter in the winter months. Because space and funding for these programs is limited, however, shelter caseworkers often do not do much outreach, and shelter residents may lose opportunities to apply for or receive new housing programs simply because they are not at the shelter when applications are being accepted.

> *Ethnographer: What about some of the other programs that the shelters offer?*
> *Tom (African American, 53): Yeah. It's been around... Well, you know, I missed, they just had a housing program at the shelter and if you miss them, you just got to wait until it comes on the next time.*

Fifteen out of the 40 homeless participants reported that they avoid shelters, preferring to double up in other people's apartments or stay on the streets. It is not necessary to reside in a shelter to receive services from a caseworker there, and several participants who stayed in shelters reported continuing to use caseworkers in shelters after they no longer resided there. However, those who never stayed in shelters often had no experiences working with caseworkers, and therefore had very little information regarding rental subsidies or housing programs.

While 60% of the participants received state Medicaid and 93% received food stamps during the study period, putting them in regular contact with the caseworkers from the Department of Social Services, all but one reported that their state caseworkers never referred them to other organizations or departments for services to meet other needs such as housing. Rather, according to participants, their role seems to be limited to processing applications for their particular programs and ensuring that information about clients is up to date in order to determine clients' continuing eligibility.

> *Ethnographer: Have you ever talked to the caseworker for anything since you got your benefits?*
>
> *Dave (African American, 45): No. All they want you to do is come down every three months and fill out the paperwork so you can get it for the next three months.*
>
> *Ethnographer: Okay. So, what are your current sources of income, you're getting SSI...*
>
> *Carol (African American, 38): That's it.*
>
> *Ethnographer: So, what does he [your caseworker] do, what's your relationship like with him?*
>
> *Carol: I'm a number... That's it. I'm just a number...Let me see, my nine digit number on my card, my insurance card, that's what I am to him.*
>
> *E: What do you talk to him about?*
>
> *P: Nothing at all.*

A few participants (N = 7) reported learning about housing programs or subsidies from caseworkers at inpatient or outpatient substance abuse treatment programs, or methadone maintenance, and three reported learning about programs by word of mouth from friends or acquaintances. For the vast majority, however, shelter staff were the primary referral agents to accessing information about housing programs.

Caseworker Priorities: "Creaming" Versus "Silting"

While all shelters had full time staff dedicated to helping shelter residents obtain permanent housing, staff from different shelters or even staff within the same shelter often had differing philosophies that affected how they processed clients. Some viewed their role as "referral agent," i.e., they referred their clients to organizations administering various supportive housing programs or housing subsidies for which they might be eligible as Mrs. Roberts described in an in-depth interview.

Ethnographer: Let's say you have a client here, let's say its Project Achieve, okay, and they qualify. Do you have to do all the follow-ups?

Mrs. Roberts: I don't, no. They do the follow-ups.

E: Okay, they help find the apartment?

Mrs. Roberts: I–referral. The client finds their apartment... They have a team of people, inspectors that will go out and inspect the apartment. Once I make the referral, initially I am out of it.

Those who viewed their role as mainly referral often described a process by which they referred clients whom they thought had the best chance at success as Mrs. Roberts described.

Ethnographer: Okay. Um, could you please describe a little bit the process in which you connect the client into a program?

Mrs. Roberts: We have applications for the different programs. If we don't, we get on the phone and call them up and ask them to send us a referral and we do referral letters. We do the screening here. If we know a person that is really actively using, as far as housing is concerned, we try to kind of get them on track because basically if they get into that apartment and they are actively using they are going to lose it. So we try not to set people up if they are.

Lipsky (1980) described this process as "creaming" and argued that this was one way that service providers cope with the demand for services being outmatched by resources. If resources are limited, then creaming is a rational strategy to ensure that resources are not wasted. Mrs. Roberts described the greatest challenge to her job as being the magnitude of the homeless problem and the limited resources available to confront the problem.

> *Mrs. Roberts: The barrier is that this homeless thing had gotten so bad. It's bigger than anybody has ever imagined that it could be. You're dealing with so many people that have some type of mental illness that everybody is just overwhelmed. The money is not there. The housing is not there...*
>
> *Ethnographer: People are overworked.*
>
> *Mrs. Roberts: Overworked, stress to no end.*

Other shelter staff described their roles as "caseworkers as advocates" actively working to increase the availability of affordable housing and particularly to provide housing to the chronically homeless. Staff of St. Mark's, a local shelter that had recently begun a supportive housing program, described this philosophy in a focus group interview.

> *Carla: [In] recent years we have begun as an organization to say that providing emergency shelter is really not the solution, but housing folks. And as we learn more about the national movement and the success rates of supportive housing for folks who are addicted to drugs and alcohol and most of our folks have co-occurring disorders, a lot of mental health and drug and alcohol addictions... There is virtually no housing. You know in terms of numbers, last year we served about 1000 unduplicated individuals through our shelter and housing program. Our board last year took the bold step of saying we really are going to transition out of the business of providing emergency shelter and we are going to become a Housing First agency...to house folks who are chronically homeless... When you actually are there starting to do the work, you know with the resources to provide the subsidies and case management, "Oh my word, how do we house these folks!" We've been doing it quite truthfully with Shelter Plus Care subsidies but we don't actually control those...But then we provide the case management services to keep them housed, but now we are at a place where we are really trying to create a program with best practice methods and have a much more organized system so that we really ultimately can be a model that other folks can look at and replicate.*

Because staff at this shelter explicitly wished to serve as a positive example of the Housing First model, they engaged in what one staff member jokingly referred to as the opposite of creaming, "silting." If success can be demonstrated with even the most difficult cases, then that provides a stronger justification for increasing funding for such programs.

Carla: Our goal has been "Let's look at the people who have the worst histories that nobody else will ever house," and that's really the approach we take, and we have created policies around that. You know we don't rule people out. We also feel strongly, and this is again the Housing First model, that you don't fix people first. They don't need to be fixed. They don't need to be ready. They just need to be housed and then you work from there and you work with intensive support.

These differences in priorities and mandates result in differences in the ways that caseworkers assist shelter residents in obtaining housing. Whereas Mrs. Roberts described making sure that clients were "ready" for housing by referring them to substance abuse or mental health treatment programs before working with them on housing, staff at St. Mark's insisted that sobriety was not a precondition for housing. Mike who obtained housing through this program confirmed this.

Ethnographer: What did they tell you about your eligibility for the program?

Mike (African American, 41): They told me what made me eligible was that one, I was a drug user and two, that I was chronically homeless.

E: Okay. Meaning?

P: Meaning that I was a prime candidate for one of the people looking for the program.

Silting, however, may result in another bias whereby less difficult clients, i.e. those without mental illness or chronic substance abuse, are not considered "prime candidates" for the program or actively recruited. One shelter resident in fact complained that "trouble makers" had received housing through this program while he, who always followed shelter rules, had not.

The Cost of Applying for Services

Lipsky (1980) argues that another way of rationing services is to increase the costs of applying for them. However, he also argues that increasing the cost of applying for services will only marginally affect demand, since those seeking many of the services that street level bureaucrats administer, e.g. housing subsidies, or welfare, need the services and cannot access them anywhere else. In other words, those with other options would not suffer the costs of seeking services. He argues that because of this, those seeking services from street-level bureaucrats should be considered "involuntary clients."

Shelter residents could be considered involuntary clients because they have few options to obtain affordable, permanent housing other than through housing subsidies or supportive housing programs. However, many participants in this project described weighing the costs of applying to various programs with the benefits they expected or hoped to receive. Whether or not a participant decided to apply for supportive housing or housing subsidies depended in part on their felt need. Those who were doubled-up with family members or friends, in addition to receiving less information about services, also may not have been as inclined to seek out information or apply for programs because their felt need was not as great, as described by Don who usually stayed in his girlfriend's apartment but occasionally stayed in a shelter when he had conflicts with her. He felt like he could use Section 8 because he often had difficulty paying his rent but had never applied.

Ethnographer: Let's say for example when you went to Green Shelter, did they ever talk to you about applying [for Section 8]? Did you ever apply?

Don (African American, 45): Nope. They never, those people never explained anything like that to you. I guess there's so many people coming through that they don't have time unless you go and want to have a session with them, you know, like me and you talking, and then you might find out some of that, but I didn't never get into that because I wasn't going to be following through on it because I was going home. I was just, you know, cooling off.

Other shelter residents chose not to apply because their expectations of receiving any benefits were small.

Ethnographer: Yeah, so you saw a lot of...hopelessness in the shelter?

Cindy (African American, 50): Yes, I mean, it's really, really hopeless. They feel like they have no one to help them, you know, and the people [shelter staff] that are there, they're just there for the paycheck, you know... It's like a revolving wheel that there's nothing happens, you know? You just push the paper, keep pushing the paper and people's lives...you know, it's really messed up.

Others weighed the costs of applying for housing against indignities suffered at the hands of social service or shelter staff.

Ethnographer: How's your relationship with your state worker?

Shawn (African American, 42): She's a smart ass... [b]ecause...you try to talk nice to her, she make you feel like you're an asshole for trying to be nice... She... act like....she giving it to me out of her pockets, which she is, which you all are, but she make you feel like you beneath.

Other costs include providing the paperwork that is required to apply for the programs or subsidies that can be difficult for persons who have been homeless for prolonged periods of time and may have lost many of their important documents. Other times, clients may not want to disclose some of their personal documents, such as arrest or medical records, which they find too personal to share and may question the necessity or relevance of such documents to their applications. Participants often expressed the feeling that these were capricious demands of caseworkers in order to delay or deny them access to housing programs.

Alex (Puerto Rican, 37): I tried all of them and there's only one of them that tried to help me, was St. Marks [shelter] but...She got her head up her butt right now. Not like that, that's how I feel. I went to go ask her [about an application to a supportive housing program] and she kept beating around the bush telling me... I have to do a certain amount of things. I have to be on a program of some type. I told her I'm on a Methadone program, that's why I'm ill, exhausted because the Methadone gets me tired. So I'm on Methadone then she came out with something else, oh this and that and this. I said, "If it ain't one thing, it's another thing." You know, she said, "Get this and now get this." And, I ain't want to get my police report.

Ethnographer: She needs the police report?

Alex: Yeah.

Ethnographer: What for?

Alex: Shelter Plus Care. That don't make no sense, huh?

Alex also implied that the program requirements, being in a drug treatment program, were too demanding for him. This sentiment was also express by Jennifer (white, 40 years old) who had applied but was not yet receiving Shelter Plus Care. She still fulfilled the supportive housing program requirements, however, in the hope that she would soon attain housing through the program. Her boyfriend, with whom she lived off and on, received the subsidy and she described the continuing demands of the program on his time.

I guess people just don't understand when you put, when you get this certificate, this, you know, Shelter Plus Care, you have so much to do. I don't know how anybody could have a job and go to all these meetings and meet with all these people that you have to meet with every day and do everything they requiring you to do. It's just, it's a lot.

Jennifer described having to meet with a caseworker monthly, be involved in daily sessions for her drug treatment, and attend weekly job readiness trainings. This particular program seems to have followed the Continuum of Care model, in that participating in several supportive services was mandatory and that applicants were expected to begin their participation in supportive services before they actually received their housing subsidy.

If the need for services is great enough, however, participants described being willing to put up with the costs of applying and expended a great deal of energy to get their needs met.

Ethnographer: Um, so you said you're applying for some housing subsidies right now?

Cindy: I'm trying to get an apartment so...

Ethnographer: Who are you working with?

Cindy: My case manager, she's my case manager, she was put in the Oasis House for me to get in touch with and I talk to her. I was so tired of calling and don't hear nothing and I know she gets tired of me calling and not getting any information. I mean...if something come through I told her to grab it, I don't care where it's at right now. You know, I'm in between a rock and a hard place at this moment. I can't be picky.

Long waiting lists also increase the cost of applying for housing subsidies such as Section 8. While waiting lists are determined by the amount of funding available, other practices, such as requiring that applicants make all requests in writing, increase costs even more.

Chris (Puerto Rican, 48): I applied for Section 8 and they said I'm in list number thousand and something but that's been for years and I call them. They say they don't take information on the phone. You have to do it through correspondence and stuff, so I didn't bother, but sometimes they do send you a letter once in a great while and they tell you what number you're at and stuff.

All but three of the participants who reported applying for Section 8 were homeless when they applied, either residing in shelters or doubled up with family members or friends. The likelihood that an applicant will still be living in the address listed on an application when correspondence is sent or subsidies become available is therefore very small, particularly since shelters place limits on the length of time a client can stay. When letters are sent to inform applicants that they have received a subsidy, they have a very limited time in which to accept the subsidy and find an apartment. If applicants do not respond because they never received the letter, they are placed again at the bottom of the waiting list, or their application is thrown out. Many participants reported applying for housing subsidies years before the baseline interview and having received no information regarding their applications since then. Many assumed that their applications were still in effect. Others found out that this was not the case only after they began working on their housing needs with caseworkers at other organizations.

Jennifer: [The caseworker] filled in the paperwork [for Shelter Plus Care] and she took both of ours actually [mine and her boyfriend's] and when we left the shelter, we had to leave the shelter, our time was up, she didn't, she ripped them up and that was after a year and a half. And then I went to St. Mark's [different shelter] and I thought it was still in effect so I said, "We have, our paperwork is with, you know," and she called up, she said, "No, they said they ripped them up." And I said, "What?."..."They ripped them up? It's been almost, it's been almost two years." And she said, "No, they ripped them up." So she said, "I'm sorry but we're going to have to start all over again." So, Sheila from the St. Mark's started it all over again.

State public assistance, such as food stamps and Medicaid, also require recipients to respond to correspondence requesting updated information twice a year to determine continued eligibility. Again, participants without a permanent address reported that they often did not receive their correspondence and frequently had their food stamps and medical benefits cut off temporarily until they could fill out the paper work and get them reinstated. As this could take between one to several weeks, it worsened their already precarious economic situation and decreased their chances of obtaining or maintaining stable housing.

Distributing Benefits and Sanctions: Favoritism Versus Discretion

Lipsky (1980) argues that while eligibility for public service benefits often may seem cut-and-dried, a considerable part of eligibility depends on the service providers' discretion. He further argues that the assignment of benefits or sanctions to clients is negotiated through "interpersonal strategies and implicit maneuvering." This negotiation, however, occurs in a context in which the service provider has much greater power over the definition of the situation and control over its outcome than the person seeking services.

Nine of the participants who stayed in shelters complained of "favoritism" in terms of who gets a bed at the shelter, how long they are allowed to stay, and who gets help with services. Sometimes this favoritism seemed based on past personal relationships that staff had with residents.

Jose (Puerto Rican, 42): Well the staff, you could have five staff members and I don't know how they do it. They tweak allotment, but there's supposed to be some kind of list, but I think it is favoritism, you know what I'm saying. Somebody that they like, they just give them a bed, you know what I'm saying. I don't think it's by a list. I think that it basically by who knows who.

Ethnographer: How would you describe the staff there?

Jose: Um, they come from my world, you know the street world, most of them. Some of them are convicts from say, you know from the same type of, how would I say it, environment I came from. Some are homeless...so they been there. Some of them are more favored towards others.

E: What brings that favoritism about?

Jose: I think um becomes from a staff-homeless person relationship. It could stem from knowing him before he came there and you know, that sort of thing.

This type of favoritism was confirmed by a shelter resident who reported having benefited from it.

Ethnographer: Why were you able to stay at the [shelter] for a while?

Dave (African American, 53): Because I really, I got along with the staff real good. Some of the staff is friends of mine, I grew up with them, you know, so they, you know I got along with them real good so I was able to stay down there. I didn't have to go out in the morning, you know, I'd stay there all day if I want to, and they let me leave on the weekends because I still was seeing my ex, so I'd go stay with her on the weekends and come back Mondays.

Other times, participants perceived that this favoritism was based on staff members' judgment of residents' deservingness.

Jose: Yeah. The favoritism. Cause certain individuals get everything. Like a comment that one of the staff said yesterday was pretty raw, upset me. He says,... "I look out for only the people that work."

Shelter staff also reported sometimes extending shelter stays for certain residents, which they described as reflecting staff discretion and program flexibility to meet client needs. They reported that they based these decisions on how "compliant" clients were to their treatment plan, or how actively they were trying to work on their goals. Residents could also have their stay shortened if they failed to follow shelter rules.

Mrs. Roberts: As long as they are working towards a goal we don't have a problem of continuing with them. It is when they are not following through on the things that they need to do that we will terminate them. Getting into a physical fight with someone would cause them to be terminated immediately...It is determined through the coordinators here. Um, we call it the coordinators team because we do work together. We seek advice from one another so that we, it won't be a one-person thing, decision, and so we do have to discuss our clients.

Mr. Green: Upon an individual's entry into a shelter, initially they have approximately 97 days to be in the shelter. Within those 97 days hopefully they're working with a case manager and they are compliant with the service plan, agreed upon service plan between the client and the case manager. [If] they are just about ready to receive certain entitlements or assistance with employment, whatever the case may be, there is a possibility for an extension to be given to an individual as long as they are compliant with the service plan that was agreed upon.

Some participants reported complying with program plans in order to extend their shelter stay while waiting for more permanent housing. Others feared being kicked out of the shelter because they were not able to comply with all the requirements.

Ethnographer: So they have like a time limit for people if they want to stay there?

Mary (white, 38): They do. Usually it is 30 days, but they have counselors there so if you are doing something, you know, about it, you know, obviously they will let you stay there. They let me stay so long because I was scheduled for a certain date to go into the Christian [drug treatment] program I was going to in New Hampshire. So that's why they let me stay and because I was in the Methadone Clinic, so I really couldn't go back to my parents, you know, because there wasn't a clinic around there.

Maribel (Puerto Rican, 38): Yeah, but it's just, I was just telling you, it's probably going to take me a damn year before I can be able to find a damn job. I said, "What are you crazy?" I can't wait that long, you know, because sooner or later I got to start paying rent where I'm at or they gonna kick me out, you know? And my case manager is pushing me like, I feel that...what I'm doing is not enough for her.

Ethnographer: Does Maria offer you any suggestions on where to go or is she just...

> *Maribel: Yeah. She offered me, she went with me too. She's pretty good, you know, in helping me out and helping me do my paperwork and something. Go here, go there, but what I mean is that, it's not enough for her...I feel like she thinks I'm out there doing whatever I want to do and not doing what I'm supposed to be doing, you know? I see her once a week, every Wednesday, which tomorrow I'll be seeing her and showing her what I have done since the last visit that I had spoke to her. I have these papers that I have to sign when I go looking for work. Anywhere I go I got to put the name, the date...of the place and try to get a signature and try to get the phone number and all this, so I can show her that I am doing this, you know?*

The decision about whether or not to extend a resident's stay at a shelter can have profound effects on whether he or she is able to access more permanent housing through subsidies or supportive housing programs. As mentioned above, most people access services through caseworkers. In addition, changing address, as residents are forced to do if their shelter stay is not extended, makes it more difficult to follow through on applications for housing subsidies.

Conclusion

The data presented in this paper illustrate that unofficial policy is as important in understanding drug users' vulnerability to homelessness and housing instability as official policy. In addition to the limits imposed by eligibility criteria and under-funding of housing subsidies, both housing caseworkers and drug users described a number of mechanisms that limit drug users' access to housing information and services. These include limited outreach regarding housing programs and subsidies to the homeless or marginally housed who avoid shelters. In addition, caseworkers prioritized clients in order to make decisions about how best to expend their limited resources and energy, "creaming" versus "silting." Another way that caseworkers rationed services is by increasing the costs of applying for them. Drug using participants described bureaucratic red tape and being treated disrespectfully by caseworkers. Finally, housing caseworkers are able to exercise considerable discretion when processing applications and serving their clients. This discretion was perceived as "favoritism" by the drug using participants interviewed in this project, and as program flexibility by caseworkers. Housing caseworkers and advocates act as "street level bureaucrats" and have developed these mechanisms as rational ways of coping with the limited resources available to perform their jobs.

Unofficial policies that are used by caseworkers to ration scarce resources help explain the relationship between structural factors (the lack of affordable housing and under-funding of subsidized housing) and personal vulnerabilities (drug abuse, arrest and mental illness) that are alternatively hypothesized to cause homelessness. Housing caseworkers perform their jobs within the constraints of the larger socio-political context operating within the United States. The Personal Responsibility and Work Opportunity Reconciliation Act (PRWORA) and the budget flat lining or cutting of many federally funded housing programs has created an increased scarcity of resources to those whose job it is to provide permanent housing to the homeless and precariously housed urban poor. At the same time, rhetoric used to justify the imposition of severe time limits on lifetime welfare benefits focus on the pernicious effects of welfare on the individual, breeding dependency and sexual immorality, and the return to "personal responsibility" and independence that revoking benefits enforced [26,37]. Drug users are particularly vilified within this system, as they constitute the undeserving poor, who have only their selfish consumption to blame for their poverty [26] and specifically targeted in official housing policy in the "One Strike Law" that bans drug users' and their families from receiving federally subsidized housing.

The service providers interviewed in this project, and whom homeless drug users must petition in order to access housing subsidies and supportive housing programs, use these same discourses to understand the reasons poor people are seeking services, and the kind of help to which they are entitled. Some saw their jobs as reforming the individuals seeking their services by providing job training assistance, mental health or drug treatment services; at the same time they blamed some clients for their homelessness because of their drug addictions, or poor work ethics. However, the passing of PRWORA also created a drastic decrease in services as state welfare offices were expected to empty welfare rolls and return recipients to work as soon as possible. Many state welfare departments disqualified welfare recipients for relatively minor infractions [38]. Like has been seen in other states [37], Connecticut Department of Social Services (DSS) caseworkers were pressured to ration services to the greatest extent possible. In the current political climate, therefore, it is hardly surprising that only one of the participants in this project reported being referred to other needed services by their DSS caseworkers, and nearly all reported having food stamps and state medical benefits discontinued for not filling out the proper re-determination paper work.

Housing caseworkers and advocates who worked in the shelters also operated in this political context. Many described clients as compliant or non-compliant, and used their discretion in order to determine who received beds on any particular night, who would be allowed to stay for an extended period at the shelter, and to whom they would devote their energy to try to help exit homelessness. The compliant homeless included those who were actively working on their "problems" by entering drug treatment or searching for a job, while the non-compliant included those who refused to comply with shelter rules or program requirements. Homeless drug users described interactions with caseworkers in overwhelmingly negative terms. They saw caseworkers as showing favoritism to some and felt that their demands of clients were unreasonable and capricious. Similar to results in this study, research on the health and social service needs of HIV infected persons has found that these persons had overwhelmingly negative experiences with service providers and case managers [39-41].

Shelter staff are also constrained by budget cuts that limit the resources available to help their clients, but have even less control than low-level DSS or HUD employees in determining who gets access to these. Shelter caseworkers used different strategies to manage their jobs under these difficult circumstances. First, shelter staff expended little energy in outreach to homeless who slept on the street or doubled up with family and friends. This is a rational strategy considering that there were too few resources to assist even those shelter residents who actively sought their services in obtaining permanent housing. Another strategy used is similar to Lipsky's "creaming" as shelter caseworkers expended time and energy assisting clients whom they thought had the greatest likelihood of success. Those deemed as having the best chance of success were those who were employed, and therefore less likely to have been incarcerated, while those who deemed unlikely to succeed included those who were actively using drugs, or with a serious mental illness. The final strategy, following the Housing First model, was "silting," in which shelter caseworkers attempted to house those deemed to be the most difficult cases, the long term homeless including chronic substance abusers, and persons with significant histories of mental illness and incarceration. By showing the success and cost-effectiveness of housing "pathological" individuals, they resisted dominant conservative discourses and advocated for expanding services to these individuals.

There are many ways of lessening the impact of the unofficial policies that serve to limit drug users' access to housing and other services. Increasing outreach to those in need of housing services who do not reside in shelters would improve access to information and housing programs to homeless persons who avoid shelters. Other valuable changes could decrease the costs of applying for services. For example, shelter staff and DSS caseworkers could receive on-going training regarding communication skills and a "customer service" approach to clients to address the lack of respect perceived by some participants. These trainings could explore and challenge caseworkers' implicit attitudes about homeless drug users and would be particularly important for organizations whose mission is to provide supportive housing to the chronically homeless. Alternative methods for the homeless to inquire about the status of their applications for housing subsidies or welfare benefits other than by mail could further decrease the costs of applying. Finally, caseworker discretion in distributing sanctions and benefits could be minimized by formalizing criteria by which decisions are made regarding extending a shelter resident's stay or who receives housing services. Improved communication regarding decision-making criteria might decrease shelter residents' perceptions of staff favoritism.

These mechanisms, however, are unlikely to be effective without devotion of significant resources and political will to solving the housing crisis. Lipsky (1988) argues that street level bureaucrats effectively become policy makers as they implement policy. Official policy, however, imposes constraints on caseworkers' ability to perform their job by defining the amount of resources available. Housing caseworkers have little incentive or power to eliminate barriers to drug users' access to housing, information and services in the current political economy. That an alternative model exists in the Housing First Model is a hopeful sign and a challenge to dominant discourse about impoverished drug users. Such challenges need to be continued and expanded in order to find solutions for chronically homeless substance abusers.

This paper is one of the first to explore how unofficial policy limits drug users' access to housing, information and service. Qualitative research is particularly well suited to explore processes such as the mechanisms housing caseworkers and advocates use to ration services. This study is additionally strengthened by its use of in-depth interviews with a larger sample of drug users than those typically included in qualitative research, the inclusion of drug users in various housing situations, the high proportion of women, and its longitudinal design. Limitations to the study include the small number of housing service providers interviewed and the lack of inclusion of non-drug using low-income, homeless or marginally housed participants. Additional qualitative and quantitative research is needed to explore ways that unofficial policy limits drug users' and non-drug users' access to housing defined as the sources of information about, perceived eligibility, application for and receipt and denial of housing programs and subsidies, welfare and social services.

Competing Interests

The author(s) declare that they have no competing interests.

Authors' Contributions

JDG conceived and designed the project, participated in the analysis and wrote the draft of the paper. MC participated in in-depth interviews, analysis of qualitative data, and helped coordinate the project. HH participated in in-depth interviews and helped in data analysis. AMC was involved in data analysis. MW participated in the design and analysis of data. All authors read and approved the final manuscript.

Acknowledgements

This paper was supported by a grant from the National Institute on Drug Abuse, DA018607.

References

1. Baum AS, Burnes DW: A Nation in Denial: the truth about homelessness. Boulder, CO, Westview Press; 1993.
2. Lamb HR: Deinstitutionalization and the homeless mentally ill. Hospital and Community Psychiatry 1984, 35:153–170.
3. Royse D, Leukefeld C, Logan TK, Dennis M, Wechsberg W, Hoffman J, Cottler L, Inciardi J: Homelessness and gender in out-of-treatment drug users. American Journal of Drug and Alcohol Abuse 2000, 26:283–296.
4. Fischer PJ, Breakey WR: The epidemiology of alcohol, drug, and mental disorders among homeless persons. American Psychologist 1991, 46:1115–1128.
5. Stahler GJ, Shipley TE, Bartelt D, Westcott D, Griffith E, Shandler I: Retention issues in treating homeless polydrug users: Philadelphia. Alcoholism Treatment Quarterly 1993, 10:201–215.
6. Dennis JL, Bray RM, Iachan R: Drug use and homelessness. In Comparative epidemiology: Examples from the DC Metropolitan Area Drug Study. Edited by: Bray RM and Marsden ME. San Francisco, Sage; 1998:79–123.
7. Lehman AF, Cordrey DS: Prevalence of alcohol, drug and mental disorder among the homeless: one more time. Contemp Drug Problems 1993, 355–386.
8. Johnson TP, Freels SA, Parsons JA, Vangeest JB: Substance abuse and homelessness: social selection or social adaptation? Addiction 1997, 92:437–445.

9. Spinner GF, Leaf PJ: Homelessness and drug abuse in New Haven. Hospital and Community Psychiatry 1992, 43:166–168.
10. Winkleby MA, Rockhill B, Jatulis D, Fortman SP: The medical origins of homelessness. American Journal of Public Health 1992, 82:1395–1398.
11. Blau J: The visible poor: homelessness in the U.S. New York, Oxford University Press; 1992.
12. Harrington M: The new American poverty. New York, Penguin Books; 1984.
13. Booth BM, Sullivan G, Koegel P, Burnam A: Vulnerability factors for homelessness associated with substance dependence in a community sample of homeless adults. Am J Drug Alcohol Abuse 2002, 28:429–452.
14. Koegel P, Burnam MA: Getting nowhere: homeless people, aimless policy. In Urban America–Policy choices for Los Angeles and the nation. Edited by: Steinberg J, Lyon D and Vaiana M. Santa Monica, CA, RAND; 1992.
15. Burt MR: Over the edge: the growth of homelessness in the 1980s. New York, Russel Sage Foundation; 1992.
16. Cohen CI, Thompson KS: Homeless mentally ill or mentally ill homeless? Am J Psychiatry 1992, 149:816–823.
17. Shinn M: Homelessness. What is a psychologist to do? Am J Community Psychology 1992, 20:1–24.
18. Koegel P, Melamid E, Burnam MA: Childhood risk factors for homelessness among homeless adults. Am J Public Health 1995, 85:1642–1649.
19. Crane J, Quirk K, van der Straten A: "Come back when you're dying":–the commodification of AIDS among California's urban poor. Social Science & Medicine 2002, 55:1115–1127.
20. Anderson TL, Shannon C, Schyb L, Goldstein P: Welfare reform and housing: assessing the impact to substance abusers. Journal of Drug Issues 2002, 32:265–296.
21. Drug Policy Alliance: State of the states: drug policy reforms: 1996–2002. New York, ; 2003.
22. Lipsky M: Street-level bureaucracy: dilemmas of the individual in public services. New York, Russel Sage Foundation; 1980.
23. Zlotnick C, Tam T, Robertson MJ: Disaffiliation, substance use, and exiting homelessness. Substance Use & Misuse 2003, 38:577–599.
24. Nyamathi A, Leake B, Keenan C, Gelberg L: Type of social support among homeless women: Its impact on psychosocial resources, health and health behaviors, and use of health services. Nurs Res 2000, 49:318–326.

25. Nwakeze PC, Magura S, Rosenblum A, Joseph H: Homelessness, Substance Misuse, and Access to Public Entitlements in a Soup Kitchen Population. Substance Use & Misuse 2003, 38:645–668.
26. Dohan D, Schmidt L, Henderson S: From enabling to bootstrapping: welfare workers' views of substance abuse and welfare reform. Contemp Drug Problems 2005, 32:429–455.
27. Hopper K, Barrow S: Two genealogies of supported housing and their implications for outcome assessment. Psychiatric Services 2003, 54:50–54.
28. Hopper K, Jost J, Hay T, Welber S, Haugland G: Homelessness, mental illness and the institutional circuit. Psychiatric Services 1997, 48:659–665.
29. Srebnik D, Livingston J, Gordon L, King D: Housing choice an community success for individuals with serious and persistent mental illness. Community Mental Health Journal 1995, 31:139–152.
30. Tsemberis S, Gulcur L, Nakae M: Housing first, consumer choice, and harm reduction for homeless individuals with a dual diagnosis. American Journal of Public Health 2004, 94:651–656.
31. Andersen A, University of Pennsylvania Health System DP Center for Mental Health Policy and Services Research, Sherwood KE: Final Program Evaluation Report. New Haven, Ct, Connecticut Supportive Housing Demonstration Program; 2002.
32. Dickson-Gomez J, Weeks M, Martinez M, Radda K: Reciprocity and Exploitation: social dynamics in private drug use sites. Journal of Drug Issues 2003, in review.
33. Weeks MR, Clair S, Singer M, Radda K, Schensul JJ, Wilson DS, Martinez M, Scott G, Knight G: High risk drug use sites, meaning and practice: Implications for AIDS prevention. J Drug Issues 2001, 31:781–808.
34. Weeks M, Dickson-Gomez J, Mosack K, Convey M, Martinez M, Clair S: The Risk Avoidance Partnership: Training active drug users as Peer Health Advocates. Journal of Drug Issues 2006, 36:541–570.
35. Weeks M, Singer M, Schensul JJ, Jia Z, Grier M: Project COPE: Preventing AIDS among injection drug users and their sex partners: descriptive data report. In Institute for Community Research. Hartford, CT; 1991.
36. Muhr T: Atlas-ti: The knowledge workbench. 2000.
37. Kingfisher C: Producing disunity: the constraints and incitements of welfare work. In The new poverty studies: The ethnography of power, politics, and impoverished people in the United States. Edited by: Goode J and Maskovsky J. New York, New York University Press; 2001.

38. Piven FF: Welfare reform and the economic and cultural reconstruction of low wage labor markets. In The new poverty studies: The ethnography of power, politics, and impoverished people in the United States. Edited by: Goode J and Maskovsky J. New York, New York University Press; 2001.
39. Seals BF, Sowell RL, Demi AS, Moneyham L, Cohen L, Guillory J: Falling through the cracks: Social service concerns of women infected with HIV. Qualitative Health Research 1995, 5:496–515.
40. Sowell RL, Phillips KD, Seals BF, Julious CH, Rush C, Spruill LK: Social service and case management needs of HIV-infected persons upon release from prison/jail. Lippincott's Case Management 2001, 6:157–168.
41. Sowell RL, Seals BF, Moneyham L, Guillory J, Demi AS, Cohen L: Barriers to health-seeking behavior for women infected with HIV. Nursing Connections 1996, 9:5–17.

Mental Health Symptoms in Relation to Socio-Economic Conditions and Lifestyle Factors – A Population-Based Study in Sweden

Anu Molarius, Kenneth Berglund, Charli Eriksson, Hans G Eriksson, Margareta Lindén-Boström, Eva Nordström, Carina Persson, Lotta Sahlqvist, Bengt Starrin and Berit Ydreborg

ABSTRACT

Background

Poor mental health has large social and economic consequences both for the individual and society. In Sweden, the prevalence of mental health symptoms has increased since the beginning of the 1990s. There is a need for a better understanding of the area for planning preventive activities and health care.

Methods

The study is based on a postal survey questionnaire sent to a random sample of men and women aged 18–84 years in 2004. The overall response rate was 64%. The area investigated covers 55 municipalities with about one million inhabitants in central part of Sweden. The study population includes 42,448 respondents. Mental health was measured with self-reported symptoms of anxiety/depression (EQ-5D, 5th question). The association between socio-economic conditions, lifestyle factors and mental health symptoms was investigated using multivariate multinomial logistic regression models.

Results

About 40% of women and 30% of men reported that they were moderately or extremely anxious or depressed. Younger subjects reported poorer mental health than older subjects, the best mental health was found at ages 65–74 years.

Factors that were strongly and independently related to mental health symptoms were poor social support, experiences of being belittled, employment status (receiving a disability pension and unemployment), economic hardship, critical life events, and functional disability. A strong association was also found between how burdensome domestic work was experienced and anxiety/depression. This was true for both men and women. Educational level was not associated with mental health symptoms.

Of lifestyle factors, physical inactivity, underweight and risk consumption of alcohol were independently associated with mental health symptoms.

Conclusion

Our results support the notion that a ground for good mental health includes balance in social relations, in domestic work and in employment as well as in personal economy both among men and women. In addition, physical inactivity, underweight and risk consumption of alcohol are associated with mental health symptoms independent of socio-economic factors.

Background

Mental health is an important part of public health. According to the Swedish national public health report [1] between 20 and 40 percent of the general population suffer from poor mental health—everything from severe psychiatric disorders such as psychosis to milder mental health symptoms such as nervousness, anxiety or sleeping problems. Whereas the most severe psychiatric disorders have not increased in the population in Sweden during the last decades, there has been an increase in the prevalence of mental health symptoms since the beginning of the 1990s. Poor mental health has large economic and social consequences both for the individual and society. The costs to society for health care, sickness absence, disability pension and loss of production due to poor mental health were estimated to 50,000 million crowns in Sweden in 1997 [2].

Results from previous studies show strong associations between mental health and e.g. social relations, income, working conditions and critical life events [3-7]. In general, persons with low socio-economic status have poorer mental health than persons with high socio-economic status [8]. Some lifestyle factors, such as physical activity [9,10], alcohol consumption [11,12] and obesity [13] have also been found to be related with mental health. In addition, domestic work has been found to be associated with mental well-being among women [14]. There is a need for a better understanding of these associations in order to plan preventive activities and health care.

The aim of the present study was to estimate the prevalence of self-reported mental health symptoms among men and women in different age groups in the general population and to disentangle the associations between socio-economic conditions, lifestyle factors and mental health symptoms. As a starting point, we used a model of mental health indicators which has been established in a working group in the European Union [3], and which includes e.g. social relations, economic factors, working conditions and critical life events. We extended the model by including domestic work and lifestyle factors in the study.

Methods

The study is based on a postal survey questionnaire sent to a random sample of men and women aged 18–84 years in autumn 2004. The aim of the survey was to investigate the health status, lifestyle factors and living conditions as well as health care use in the population. The sampling was random at individual level and stratified by gender, age group, county and municipality. The data collection was completed after two postal reminders. The overall response rate was 64 percent. The area investigated covers 55 municipalities in five counties with about one million inhabitants in central part of Sweden. The study population includes 42,448 respondents.

Mental health symptoms were measured with a question about anxiety/depression (EQ-5D, 5th question). EQ-5D [15] is a standardised instrument including five questions that measure health related quality of life. The 5th question represents mental health and is as follows. "Please indicate which statements best describe your own health state today: Anxiety/Depression", with answer options I am not anxious or depressed, I am moderately anxious or depressed and I am extremely anxious or depressed.

Socio-Economic Conditions

Educational level was obtained through record linkage to information from a national education register and was categorised into three classes: low (elementary school), medium (upper secondary school), and high (at least 3 years of university or corresponding education). Country of origin was obtained by record linkage to a national population register. The respondents were categorised into those born in Sweden, in other Nordic countries, in other European countries and outside Europe. Family status was obtained from a survey question and categorised into living alone, living with partner, living with partner and children, single parent and other.

Employment status was derived from a survey question about whether the respondent was employed, self-employed, student, on parental leave, unemployed, working at home, on disability pension, retired or other. Economic hardship was assessed by asking whether the respondent had had problems with paying running bills during the last 12 months (no problems; yes, during 1–2 months; yes, during 3–12 months).

Social support was assessed with the question "Do you have any persons in your surrounding you can get support from in case of emotional crises or problems?" with the answer options yes, definitely; yes, probably; probably not and definitely not. The participants were also asked whether they had experienced that someone had belittled them during the last three months. The answer categories were never, once or twice, and several times during the last three months.

There were two questions about domestic work. The first asked how many hours per week the respondent spent working at home that was not paid work (e.g. taking care of children, nursing relatives, buying the groceries, cooking, paying the bills, washing the laundry, cleaning, taking care of a car, house or garden). The second question asked how often the respondent experienced domestic work as burdensome (all or most of the time, sometimes, seldom, never).

The statement: "One can trust the people living in this neighbourhood" was used to evaluate neighbourhood social cohesion where agreement was coded as good, partial disagreement as less good and total disagreement as poor social cohesion. Participation in associations was asked with a question whether the respondent was an active member in an association (trade union, political party, nature/environmental association, sports club, pensioners association, religious association, cultural association, administrative board, other).

Physical environment was derived from a question: “How often do you have disturbance in or around your house from the following sources?” with the alternatives: noise from outside, exhaust from outside, disturbing industry, draught and cold, disturbing neighbours, bad smell, poor quality of drinking water, littered environment, damage or graffiti and other disturbance with the options often, sometimes, seldom and never. Option never was then coded as 0, sometimes or seldom were coded as 1 and often as 2 for each of the disturbances. If the sum was 0–2 the physical environment was coded as good, 3–5 was coded as less good and 6 or more was coded as poor physical environment.

The respondent was defined as functionally disable if she/he needed help on a daily basis due to functional disability or illness. Critical life events during the last two years (death of a near relative, own or a relative's severe illness, separated from a spouse or a partner, being laid off from work, other critical life event) were asked and dichotomised into no or at least one event.

Lifestyle Factors

Physical activity was measured with the question: “How much do you exercise physically in your leisure time?” with the options little exercise (walking, bicycling or other light exercise less than 2 hours a week), moderate exercise (walking, bicycling or other light exercise more than 2 hours a week), moderate regular exercise (exercising 1–2 times a week at least for half an hour at a time in jogging, playing tennis, bicycling, exercising at a gym or other moderate exercise that makes one to sweat) and vigorous exercise and training (exercising or competing at least 3 times a week at least for half an hour at a time in team sports, jogging, playing tennis, swimming or other intensive physical activity). The two middle categories were combined into moderate exercise.

Smoking habits and snuff use were derived from the questionnaire, combined and dichotomised into any cigarette smoking or snuff use daily and not daily. Alcohol consumption was measured using the first three questions in the WHO instrument AUDIT (Alcohol Use Disorders Identification Test). These three questions measure the frequency and quantity of alcohol consumption and relate to risk consumption of alcohol [16].

Relative weight was measured by using body mass index (BMI). BMI was calculated from self-reported weight and height as weight divided with height squared (kg/m^2). The participants were categorised according to the WHO guidelines

[17] as underweight when BMI was lower than 18.5 kg/m^2, normal weight when BMI was between 18.5 and 24.9 kg/m2, overweight when BMI was between 25 and 29.9 kg/m^2, and obese when BMI was equal to or over 30 kg/m^2.

The respondents gave their informed consent to use the national register data by answering the questionnaire. The personal identification numbers were deleted directly after the record linkage with the national registers and the survey data are thus anonymous. The survey was approved by the boards of the five county councils and the confidentiality of the data is assured under the Swedish law.

Statistical Analyses

The prevalence of mental health symptoms (using EQ-5D, 5th question) is reported by gender and age group (Figure 1). The association between socio-economic conditions, lifestyle factors and mental health symptoms was investigated using multinomial logistic regression models. The results are reported as odds ratios (OR) and 95 percent confidence intervals (95% CI) for being extremely or moderately anxious/depressed, respectively, when adjusting for all the other variables in the model (Table 1). The category of not anxious/depressed was the constant category. Since the associations between the studied factors and mental health symptoms were fairly similar in men and women, the combined analyses are reported in this paper, adjusted for gender. Some differences in these associations between men and women are, however, commented in the text. Age was not independently associated with mental health symptoms and was therefore not included in the fully adjusted model (Table 1). The variables that were statistically significantly related to anxiety/depression in univariate analyses were included in the fully adjusted model.

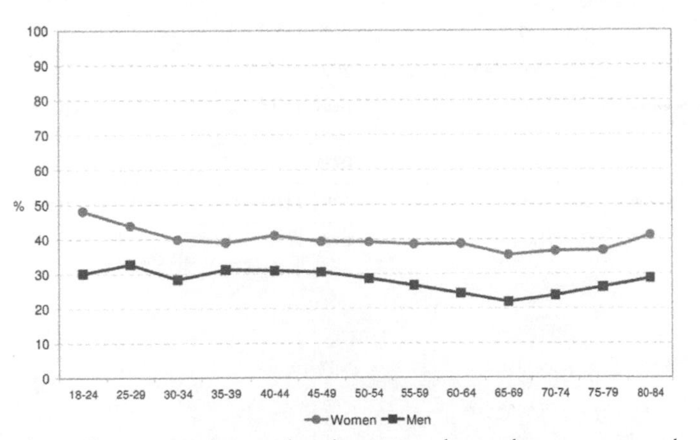

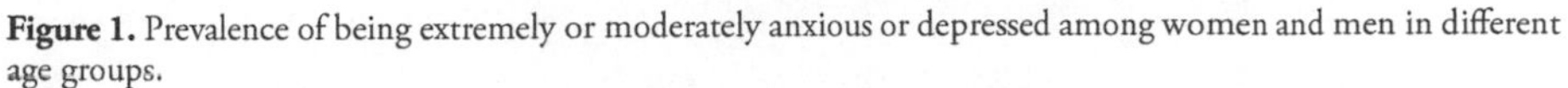

Figure 1. Prevalence of being extremely or moderately anxious or depressed among women and men in different age groups.

Table 1. Odds ratios (with 95% confidence intervals) for being extremely or moderately anxious or depressed.

Variable	Category	N	OR for extremely anxious/ depressed	OR for moderately anxious/ depressed
Socio-economic conditions				
Social support	No	661	5.3 (3.7, 7.4)	1.8 (1.5, 2.1)
	Probably not	858	6.6 (4.8, 9.0)	3.0 (2.6, 3.6)
	Probably yes	7314	2.3 (1.9, 2.8)	1.6 (1.5, 1.7)
	Yes, definitely (ref.)	25812	1	1
Being belittled (last 3 months)	Several times	1097	11.5 (8.8, 15.1)	3.6 (3.1, 4.2)
	Once or twice	6786	3.3 (2.7, 4.0)	2.1 (2.0, 2.2)
	Never (ref.)	26762	1	1
Employment status	Disability pensioner	1858	8.5 (6.5, 11.0)	2.9 (2.6, 3.3)
	Unemployed	1697	2.9 (2.2, 4.0)	1.6 (1.4, 1.8)
	Student	2068	2.0 (1.4, 2.8)	1.3 (1.1, 1.4)
	Retired	8516	1.4 (1.1, 1.9)	1.3 (1.2, 1.4)
	Self-employed	1953	1.1 (0.7, 1.9)	1.0 (0.9, 1.1)
	Parental leave	672	1.1 (0.6, 2.2)	0.8 (0.6, 0.9)
	Other	1119	4.3 (3.1, 6.1)	1.5 (1.3, 1.8)
	Employed (ref.)	16762	1	1
Economic hardship (last 12 months)	Yes, 3–12 months	2042	3.1 (2.4, 3.9)	1.9 (1.7, 2.1)
	Yes, 1–2 months	3242	1.5 (1.2, 1.9)	1.4 (1.3, 1.5)
	No (ref.)	29361	1	1
Domestic work burdensome	All the time	2540	7.2 (5.4, 9.7)	3.0 (2.7, 3.4)
	Sometimes	13994	2.3 (1.8, 3.0)	1.9 (1.8, 2.0)
	Seldom	8538	1.1 (0.8, 1.5)	1.3 (1.2, 1.4)
	Never (ref.)	9573	1	1
Critical life events (last 2 years)	Yes	19006	2.2 (1.8, 2.6)	1.4 (1.4, 1.5)
	No (ref.)	15639	1	1
Functional disability	Yes	1684	3.4 (2.6, 4.4)	2.0 (1.8, 2.3)
	No (ref.)	32961	1	1

Table 1 *(Continued)*

Family status	Living with partner	13845	1.3 (1.0, 1.6)	1.2 (1.1, 1.3)
	Living alone	5868	2.1 (1.6, 2.6)	1.4 (1.3, 1.5)
	Single parent	2126	1.3 (1.0, 1.7)	1.2 (1.1, 1.3)
	Other	1693	1.5 (1.0, 2.2)	1.3 (1.2, 1.5)
	Living with partner and children (ref.)	11113	1	1
Physical environment	Poor	4081	1.5 (1.2, 1.8)	1.5 (1.4, 1.6)
	Less good	10080	1.2 (1.0, 1.5)	1.3 (1.2, 1.3)
	Good (ref.)	20484	1	1
Participation in associations	No	21386	1.9 (1.5, 2.3)	1.3 (1.3, 1.4)
	Yes (ref.)	13259	1	1
Neighbourhood social cohesion	Poor	498	1.2 (0.8, 1.9)	1.3 (1.0, 1.6)
	Less good	1637	1.4 (1.1, 1.9)	1.2 (1.1, 1.4)
	Good (ref.)	32510	1	1
Background factors				
Gender	Female	18443	1.3 (1.1, 1.6)	1.5 (1.4, 1.6)
	Male (ref.)	16202	1	1
Country of birth	Outside Europe	1025	2.0 (1.4, 2.8)	1.3 (1.1, 1.5)
	Other European country	923	2.0 (1.3, 2.9)	1.3 (1.1, 1.5)
	Other Nordic country	1694	0.8 (0.6, 1.2)	1.0 (0.9, 1.2)
	Sweden (ref.)	31003	1	1
Lifestyle factors				
Physical activity	Inactive (<2 h/week)	6070	2.2 (1.6, 3.1)	1.6 (1.4, 1.7)
	Moderate	23897	1.4 (1.0, 1.9)	1.3 (1.2, 1.4)
	Vigorous (ref.)	4678	1	1
Smoking or snuff use	Daily	8346	1.4 (1.2, 1.7)	1.2 (1.1, 1.2)
	Not daily (ref.)	26299	1	1
Risk consumption of alcohol	Yes	2078	1.6 (1.2, 2.1)	1.4 (1.3, 1.6)
	No (ref.)	32567	1	1
	Normal weight (ref.)	16727	1	1
	Overweight (25–29.9)	12869	0.9 (0.8, 1.1)	1.0 (1.0, 1.1)
	Obese (>30)	4583	0.9 (0.7, 1.2)	1.0 (1.0, 1.1)

Multinomial logistic regression, all independent variables are included in the same model, age range 18–84 years.

Results

About 40 percent of women and 30 percent of men reported that they were moderately or extremely anxious or depressed. The prevalence of mental health symptoms was higher among younger than older subjects. The best mental health was found at ages 65–74 years (Figure 1).

Of the 34,645 subjects aged 18–84 years who answered all the questions included in the multiple multinomial logistic regression analysis, 10,697 (31 percent) reported that they were moderately anxious or depressed, whereas 672 (2 percent) reported that they were extremely anxious or depressed.

Factors that were strongly and independently related with anxiety/depression were poor social support, experiences of being belittled, employment status (receiving a disability pension and unemployment), economic hardship, critical life events, and functional disability (Table 1). There was no association between the number of hours spent in domestic work and mental health symptoms. Instead a strong independent association was found between how burdensome domestic work was experienced and anxiety/depression. This was true for both men and women.

Country of origin and family status were also associated with mental health symptoms (Table 1). Subjects born in other European countries and outside Europe were more often anxious or depressed than those born in Nordic countries. Persons living alone had a higher prevalence of anxiety/depression than persons living with partner and children. Single parents had a high odds ratio for being extremely anxious/depressed in the univariate analysis (OR: 3.5, 95% CI: 2.8, 4.4) but this association attenuated considerably when adjusting for socio-economic and lifestyle factors.

Participation in associations, neighbourhood social cohesion and physical environment were only slightly associated with mental health symptoms when adjusted for the other factors included in the model. Educational level and age were not independently associated with mental health symptoms.

Of lifestyle factors, physical inactivity, underweight and risk consumption of alcohol were independently associated with anxiety/depression. Underweight was associated with anxiety/depression especially among women and risk consumption of alcohol especially among men (not shown).

Discussion

Whereas the most severe psychiatric disorders, such as psychoses, have not increased in the population in Sweden during the last decades, there has been an increase in the prevalence of nervousness and anxiety since the beginning of the 1990s [1]. One possible explanation that has been mentioned is that it has become more socially accepted to tell about nervousness or anxiety. The increased premature mortality and psychiatric morbidity associated with these symptoms has, however, been relatively stable during the last ten years, indicating that self-reported severe mental health symptoms are good indicators of psychiatric morbidity [18].

Our results show that women report mental health symptoms to a larger extent than men do. It is plausible that this has to do with the position of women in society. Even though there is a relatively high equality of opportunities between genders in Sweden, women still have a high workload both at work and at home [19] and therefore also a higher level of stress hormones [20].

Young adults have a higher prevalence of mental health symptoms than older subjects do. Nearly half of women and one third of men aged 18–34 years reported that they were moderately or extremely anxious or depressed. The prevalence of mental health symptoms decreased with age until the age of 70–74 years and increased again among those over 75 years. Many factors that have been shown to be associated with mental health symptoms in the present and other studies (such as unemployment, economic hardship and being belittled) are more prevalent among younger than older subjects.

Factors Associated with Mental Health Symptoms

Social relations are in many ways important for mental health [4]. Social support is a protecting factor that acts a buffer in psychosocial crisis situations and strain [21]. Poor social support and being belittled were strongly related with mental health symptoms in the present study. Previous studies indicate that experiences of shame are associated with poor mental health for example among the unemployed [22].

Personal economy had also a strong association with mental health symptoms. Subjects with economic problems had a higher prevalence of anxiety/depression than subjects without economic problems. Previous studies have indicated that economic hardship both at present [5,23,24] and under childhood [24] is strongly associated with poor mental health.

There was no association between the number of hours spent in domestic work such as taking care of children, nursing relatives, buying the groceries, cooking, washing the laundry, cleaning etc. and mental health symptoms. Instead a strong independent association was found between how burdensome domestic work was experienced and anxiety/depression. Subjects who often or all the time experienced domestic work as burdensome had an increased prevalence of mental health symptoms. This was true as well for women as for men. Previous studies have reported domestic work as a risk factor for poor health among women, particularly in combination with work-related stress [14,20,25], whereas the association has been less often studied or found weaker among men.

Critical life events, such as death of a near relative, own or a relative's severe illness, separation from a spouse or a partner or being laid off from work, were associated with mental health symptoms in the present study. These events can be a triggering factor for poor mental health because they require a high level of psychological adaptation [21]. There is also an association between physical ill health and mental ill health [1,3]. In the present study, a factor that was strongly related to mental health symptoms was functional disability i.e. being dependent on help from others to manage everyday life.

Single parents have been found to have higher level of mental health problems than population in general [1,26]. In the present study, there was a strong crude association between being single parent and mental health symptoms. This association, however, almost disappeared when adjusted for other socio-economic conditions and lifestyle factors, suggesting that the increased level of anxiety/depression among single parents can be explained by these factors. For example, burdensome domestic work and economic hardship are more prevalent among single parents than parents living together. This should be taken into consideration when reporting differences in mental health symptoms between different family constellations. On the contrary, the association between living alone and anxiety/depression remained about the same even after the adjustment.

There was also an association between country of origin and mental health symptoms. Subjects born in other European countries and outside Europe were more often anxious or depressed than those born in Nordic countries which is in line with previous studies [1]. Women had a somewhat higher prevalence of anxiety/depression than men even when socio-economic conditions and lifestyle factors were taken into account.

Working conditions, such as high demands in combination with low control at work and job insecurity have been shown to be detrimental for health [6,7,27]. To elucidate the role of working conditions was, however, beyond the scope of the present study. Subjects who were not employed, such as disability pensioners and the unemployed, had a higher level of anxiety/depression than the employed, which is in agreement with previous studies [1,7,28].

Physical inactivity was associated with mental health symptoms in the present study. This is in line with previous studies where physical activity has been shown to have a positive effect on mental health [9,10]. Underweight subjects had a higher prevalence of mental health symptoms than normal weight subjects, especially among women, corroborating previous studies [29]. Underweight can be an effect of an eating disorder, which in turn is related to poor mental health. Contrary to previous research [13] there was, however, no association between obesity and anxiety/depression when adjusted for socio-economic and other lifestyle factors.

A high and long lasting consumption of alcohol increases the risk of alcohol related injuries, suicide, depression and anxiety [11,12]. It has been shown in national studies in Sweden that risk consumption of alcohol is related with depression and anxiety [11]. In the present study, there was a strong independent association between risk consumption of alcohol and self-reported anxiety/depression among men.

Strengths and Limitations of the Study

Since the present study is based on cross-sectional data, it is not possible to say which are causes and which are effects of mental health symptoms. In many cases the relationships are bi-directional [3,30]. For example, problems in social relations can lead to mental health symptoms, but poor mental health can also lead to problems in social relations. Economic hardship can cause anxiety or depression, but anxiety/depression can lead to economic hardship through lower income due to sickness absence or disability pension. Furthermore, burdensome domestic work can lead to mental health symptoms, but poor mental health can also lead to that one experiences domestic work as burdensome.

The response rate of the present study was 64 percent. The response rate was lower among younger than older subjects and among men compared with women. The level of education was also somewhat higher among the respondents than among the general population of the same age. Those who suffer from severe psychiatric disorders are probably underrepresented. Therefore the absolute levels of self-reported mental health symptoms should be interpreted with caution. It is, however, unlikely that the associations between mental health symptoms and other factors reported in the present study could have been explained by non-response.

A strength of the present study is that it is large and population-based. It comprises a study population of over 42,000 individuals and represents about one million inhabitants aged 18–84 years in Sweden. We could even study factors that are rare in the general population and take into account a wide range of socio-economic and lifestyle factors at the same time.

EQ-5D is an internationally validated scale of quality of life where the fifth dimension measures anxiety/depression [15,31]. Another widely used measure of mental health is GHQ-12, the twelve-item version of the General Health Questionnaire [32], which was also measured in the present study. We used EQ-5D to analyse the association between the studied socio-economic and lifestyle factors and mental health because it gives more information about the severity of mental health symptoms than using one cut-off point for GHQ-12. The results were, however, similar when using GHQ-12 instead of EQ-5D as the dependent variable, which gives further support to the findings of the study.

As a starting point of the study, we used a model of mental health indicators which has been established in a working group in the European Union [3]. It includes e.g. social relations, economic factors, working conditions and critical life events. We were, however, able to extend the model by elucidating the importance of domestic work and lifestyle factors in the same context.

Conclusion

Our results support the notion that a ground for good mental health includes balance in social relations, in domestic work, in employment as well as in personal economy. This is in line with previous studies, but adds domestic work as one of the key factors both among men and women. In addition, lifestyle factors such as physical inactivity, underweight and risk consumption of alcohol seem to be associated with mental health symptoms independent of socio-economic factors. It would be valuable to take into account all these areas of life when planning activities to prevent mental health symptoms, highly prevalent in the general population, and when promoting mental health. Furthermore, an individual will be able to better handle psychosocial crisis situations or strain if she/he possesses a wide array of protecting factors.

Competing Interests

The authors declare that they have no competing interests.

Authors' Contributions

All authors participated in acquisition of the data, design of the study and helped to draft the manuscript. AM co-ordinated the study and drafted the manuscript. HE, CP, KB and MLB performed the statistical analyses. All authors read and approved the final manuscript.

Acknowledgements

This study was funded by the County Councils of Västmanland, Uppsala, Örebro, Sörmland and Värmland.

References

1. Stefansson CG: Major public health problems–mental ill-health. In Health in Sweden–The National Public Health Report 2005. Scand J Public Health 2006, 34(Suppl 67):87–103.
2. Välfärd och valfrihet? Slutrapport från utvärderingen av 1995 års psykiatrireform [Welfare and freedom of choice? Final report from the evaluation of the 1995 psychiatry reform] Stockholm: Socialstyrelsen; 1999.
3. Korkeila J, Lehtinen V, Bijl R, Dalgard AS, Kovess V, Morgan A, Salize HJ: Establishing a set of mental health indicators for Europe. Scand J Public Health 2003, 31:451–9.
4. Cooper H, Arber S, Fee L, Ginn J: The influence of social support and social capital on health. A review and analysis of British data. London: Health Education Authority; 1999.
5. Martikainen P, Adda J, Ferrie JE, Davey Smith G, Marmot M: Effects of income and wealth on GHQ depression and poor self-rated health in white collar women and men in the Whitehall II study. J Epidemiol Community Health 2003, 57:718–23.
6. Karasek R, Theorell T: Healthy work, stress, productivity and the reconstruction of working life. New York: Basic Books; 1990.
7. Ferrie JE, Shipley MJ, Stansfield A, Marmot MG: Effects of chronic job insecurity and change in job insecurity on self-reported health, minor psychiatric morbidity, physiological measures, and health related behaviours in British civil servants: the Whitehall II study. J Epidemiol Community Health 2002, 56:450–4.
8. Dohrenwend BP: Socio-economic status (SES) and psychiatric disorders. Soc Psychiatry Psychiatr Epidemiol 1990, 25:41–7.
9. Penedo FJ, Dahn JR: Exercise and well-being: a review of mental and physical health benefits associated with physical activity. Curr Opin Psychiatr 2005, 18:89–93.
10. Hamer M, Stamakis E, Steptoe A: Dose response relationship between physical activity and mental health: The Scottish Health Survey. Br J Sports Med 2008, in press.

11. Boström G: Habits of life and health. In Health in Sweden–The National Public Health Report 2005. Scand J Public Health 2006, 34(Suppl 67):199–228.
12. Smith GW, Shevlin M: Patterns of alcohol consumption and related behaviour in Great Britain: A latent class analysis of the Alcohol Use Disorder Identification Test (AUDIT). Alcohol Alcohol 2008, 43:590–4.
13. Scott KM, Bruffaerts R, Simon GE, et al.: Obesity and mental disorders in the general population: results from the world mental health surveys. Int J Obesity 2008, 32:192–200.
14. Lindfors P, Berntsson L, Lundberg U: Total workload as related to psychological well-being and symptoms in full-time employed female and male white-collar workers. Int J Behav Med 2006, 13:131–7.
15. EuroQol Group [http://www.euroqol.org].
16. Bergman H, Källmén H, Rydberg U, Sandahl C: Tio frågor om alkohol identifierar beroendeproblem. [Ten questions about alcohol as identifier of addiction problems. Psychometric tests at an emergency psychiatric department]. Läkartidningen 1998, 95:4731–35.
17. Obesity–preventing and managing the global epidemic Report of a WHO Consultation on Obesity. Geneva 1997.
18. Ringbäck Weitoft G, Rosén M: Is perceived nervousness and anxiety a predictor of premature mortality and severe morbidity? A longitudinal follow up of the Swedish survey of living conditions. J Epidemiol Community Health 2005, 59:794–8.
19. Krantz G, Lundberg U: Workload, work stress, and sickness absence in Swedish male and female white-collar employees. Scand J Public Health 2006, 34:238–46.
20. Lundberg U, Frankenhauser M: Stress and workload of men and women in high-ranking positions. J Occup Health Psychol 1999, 4:142–51.
21. Brown GW, Harris TO: Social origins of depression. A study of psychiatric disorder in women. London: Tavistock; 1978.
22. Starrin B, Jönsson LR: The Finances-Shame Model and the Relation Between Unemployment and Health. In Unemployment and health: International and interdisciplinary perspectives. Edited by: Kieselbach T, Winefield AH, Boyd C, Anderson A. Brisbane: Australian Academic Press; 2006.
23. Weich S, Lewis G: Poverty, unemployment, and common mental disorders. BMJ 1998, 317:115–9.

24. Lahelma E, Laaksonen M, Martikainen P, Rahkonen O, Sarlio-Lähteenkorva S: Multiple measures of socioeconomic circumstances and common mental disorders. Soc Sci Med 2006, 63:1383–99.
25. Mellner C, Krantz G, Lundberg U: Symptom reporting and self-rated health among women in mid-life: the role of work characteristics and family responsibilities. Int J Behav Med 2006, 13:1–7.
26. Weitoft GR, Haglund B, Hjern A, Rosén M: Mortality, severe morbidity and injury among long-term lone mothers in Sweden. Int J Epidemiol 2002, 31:573–80.
27. Stansfield S, Candy B: Psychosocial work environment and mental health–a meta-analytic review. Scand J Work Environ Health 2006, 32:443–62.
28. Janlert U: Unemployment as a disease and diseases of the unemployed. Scand J Work Environ Health 1997, 23(Suppl 3):79–83.
29. Ali SM, Lindström M: Socioeconomic, psychosocial, behavioural, and psychological determinants of BMI among young women: differing patterns for underweight and overweight/obesity. Eur J Public Health 2006, 16:325–31.
30. Johnson JG, Cohen P, Dohrenwend BP, Link BG, Brook JS: A longitudinal investigation of social causation and social selection processes involved in the association between socioeconomic status and psychiatric disorders. J Abnorm Psychol 1999, 108:490–9.
31. Henriksson M, Burström K: Kvalitetsjusterade levnadsår och EQ-5D. En introduktion. Läkartidningen 2006, 103:1734–9.
32. McDowell I, Newell C: Psychological well-being. In Measuring health, a guide to rating scales and questionnaires. 2nd edition. Oxford: Oxford University Press; 1996:177–237.

Copyrights

distribution, and reproduction in any medium, provided the original work is properly cited.

3. © 2009 Bowen et al.; licensee BioMed Central Ltd. This is an Open Access article distributed under the terms of the Creative Commons Attribution License (http://creativecommons.org/licenses/by/2.0), which permits unrestricted use, distribution, and reproduction in any medium, provided the original work is properly cited.
4. © 2009 Nordt et al.; licensee BioMed Central Ltd. This is an Open Access article distributed under the terms of the Creative Commons Attribution License (http://creativecommons.org/licenses/by/2.0), which permits unrestricted use, distribution, and reproduction in any medium, provided the original work is properly cited.
5. © 2009 Wechsberg et al.; licensee BioMed Central Ltd. This is an Open Access article distributed under the terms of the Creative Commons Attribution License (http://creativecommons.org/licenses/by/2.0), which permits unrestricted use, distribution, and reproduction in any medium, provided the original work is properly cited.
6. © 2009 van Olphen et al.; licensee BioMed Central Ltd. This is an Open Access article distributed under the terms of the Creative Commons Attribution License (http://creativecommons.org/licenses/by/2.0), which permits unrestricted use, distribution, and reproduction in any medium, provided the original work is properly cited.
7. © 2009 Root et al. This is an open-access article distributed under the terms of the Creative Commons Attribution License, which permits unrestricted use, distribution, and reproduction in any medium, provided the original author and source are credited.
8. © 2008 Fischer et al.; licensee BioMed Central Ltd. This is an Open Access article distributed under the terms of the Creative Commons Attribution License (http://creativecommons.org/licenses/by/2.0), which permits unrestricted use, distribution, and reproduction in any medium, provided the original work is properly cited.
9. © 2009 M. van't Wout, S. van Rijn, T. Jellema, R.S. Kahn, and A. Aleman. This is an open-access article distributed under the terms of the Creative Commons Public Domain declaration which stipulates that, once placed in the public domain, this work may be freely reproduced, distributed, transmitted, modified, built upon, or otherwise used by anyone for any lawful purpose.
10. © 2008 Wilkinson et al.; licensee BioMed Central Ltd. This is an Open Access article distributed under the terms of the Creative Commons Attribution License

Index

H

I

K

L

M

N

O

P

W

X

Y